GRID COMPUTING:
A PRACTICAL GUIDE
TO TECHNOLOGY
AND APPLICATIONS

Grid Computing: A Practical Guide to Technology and Applications

Ahmar Abbas

CHARLES RIVER MEDIA, INC.
Hingham, Massachusetts

Editor: David Pallai
Production: Publishers' Design and Production Services, Inc
Cover Design: The Printed Image

CHARLES RIVER MEDIA, INC.
10 Downer Avenue
Hingham, Massachusetts 02043
781-740-0400
781-740-8816 (FAX)
info@charlesriver.com
www.charlesriver.com

This book is printed on acid-free paper.

Ahmar Abbas. *Grid Computing: A Practical Guide to Technology and Applications.*
ISBN: 1-58450-276-2

All brand names and product names mentioned in this book are trademarks or service marks of their respective companies. Any omission or misuse (of any kind) of service marks or trademarks should not be regarded as intent to infringe on the property of others. The publisher recognizes and respects all marks used by companies, manufacturers, and developers as a means to distinguish their products.

Library of Congress Cataloging-in-Publication Data
Abbas, Amhar.
   Grid computing : a practical guide to technology and applications /
Amhar Abbas.
      p. cm.
  ISBN 1-58450-276-2
  1. Computational grids (Computer systems) I. Title.
  QA76.9.C58A23 2003
  004'.36—dc22
                 2003018622

Printed in the United States of America
04 7 6 5 4 3 2

CHARLES RIVER MEDIA titles are available for site license or bulk purchase by institutions, user groups, corporations, etc. For additional information, please contact the Special Sales Department at 781-740-0400.

Requests for replacement of a defective CD-ROM must be accompanied by the original disc, your mailing address, telephone number, date of purchase and purchase price. Please state the nature of the problem, and send the information to CHARLES RIVER MEDIA, INC., 10 Downer Avenue, Hingham, Massachusetts 02043. CRM's sole obligation to the purchaser is to replace the disc, based on defective materials or faulty workmanship, but not on the operation or functionality of the product.

This book is dedicated to my late mother,
Mrs. Zenobia Abbas

Contents

Preface

We are experiencing what can only be called the *Perfect Storm* of innovation, in not just one but many of the fundamental technologies of our day. Computing systems, storage systems, and networking systems are seeing exponential growth in capacity and capabilities, while their costs are declining at an equally astonishing rate. The emergence of this new innovation-driven economics is sending shock waves through the information technology marketplace. Technologists and business managers are reassessing how much and what kinds of systems and services they should purchase and where and how these systems should be deployed.

Grid Computing offers a suite of technologies that explicitly recognizes the new economics of computing and networking and provides the tools that companies can use to drastically cut technology expenditures, increase productivity of technology assets and employees, and have a positive impact on the corporate bottom line.

Grid Computing takes collective advantage of the vast improvements in microprocessor speeds, optical communications, raw storage capacity, the World Wide Web, and the Internet that have occurred over the last few years. A set of standards and protocols are being developed that completely disaggregate current compute platforms and distribute them across a network as resources that can be called into action by any eligible user (person or machine) at any time. For example, a company with 1000 *grid-enabled* desktop PCs can utilize all of them together as one compute platform—suddenly providing it with enough computing capacity to go head-to-head with the world's 150th largest supercomputer. This tremendous capacity is being put to good use by companies in hundreds of different ways—such as accelerating drug development, processing complex financial models, animating movies, etc.

Grid Computing, like many other exciting technologies, has been developed and carefully nurtured by the collective talent and dedication of academics, researchers, and scientists the world over. Grid Computing has enabled the scientific community to attack the toughest of e-science problems including the vaunted *Grand Challenges of Science*.

Also, like many other frontier technologies, Grid Computing has made the transition from its academic and research roots to the commercial realm. It is those commercial and practical applications of Grid Computing that this work attempts to address. Readers could not ask for a more knowledgeable set of contributors, who have been carefully selected from the ranks of software houses and research organizations and who have made it their *raison d'etre* to deliver the benefits of Grid Computing to the enterprise technology community at large.

This work can be divided into three sections. The first, which includes Chapters 1, 2, and 3, defines the context surrounding information technology in businesses today. A rationale is presented on why Grid Computing is poised to make dramatic improvements to the corporate IT infrastructure.

The second section, which encompasses Chapters 4 through 15, delves into the details of Grid Computing technology. Practical considerations in implementing Grid Computing are discussed as well.

Finally, the third section, which covers Chapters 16 through 20, presents examples of grid deployments in industry and research. Specific industry sectors such as bioinformatics, financial services, and telecommunications are discussed in detail.

It is our hope that readers will consider Grid Computing an empowering technology with which they can take a *proactive role* in defining and shaping the way information technology provides their companies a competitive edge for years to come.

HOW TO USE THIS BOOK

For readers new to Grid Computing, the best way to tackle the subject is to simply read each chapter in order. However, because, it has been our intent to write each chapter as a standalone module, readers interested in, e.g., data grids, can go directly to Chapter 8, "Data Grids." Readers who want to delve deeper into Grid Computing are encouraged to refer to the references listed at the end of most chapters.

In addition, The GridBlog (*www.gridblog.com*), a Grid Computing news and analysis site, provides a vast array of resources, links, and commentary that can help readers remain up-to-date on Grid Computing technology, its marketplace, and its practical applications.

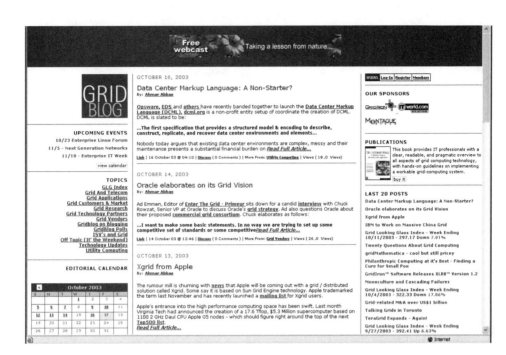

Acknowledgments

This book would not be possible without the generosity of all the contributors, who carved out time from their hectic schedules as leaders in the Grid Computing marketplace to share practical and real world knowledge of Grid Computing technology.

I'd also like to thank Fran DuCharme of Entropia, Juliette Schmidt of Platform Computing, Louise Bono of Avaki Corporation, Tim Williams of United Devices and Colleen Hagerty of Greenough Communications for all their assistance.

A great many thanks to Carl Gehrman for sharing his fantastic graphics design talent.

This book would not have been possible without the tremendous encouragement and support of my parents, sisters, and family in Pakistan, and to my in-laws in the United States.

Finally, I am deeply grateful to my wife Nabeela Khatak, who is my true partner in this and every other endeavor I undertake in life. She not only contributed her numerous talents to this book but did so while welcoming our first child, Neshmeeya, into this world.

1 | IT Infrastructure Evolution

Ahmar Abbas

Grid Technology Partners

In This Chapter

- Infrastructure Technology Ensemble
- Laws of Exponential Growth

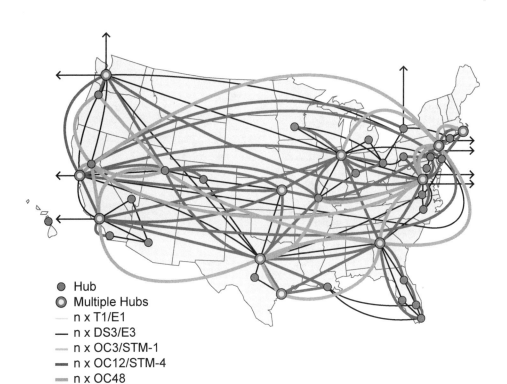

- ● Hub
- ◎ Multiple Hubs
- — n x T1/E1
- — n x DS3/E3
- — n x OC3/STM-1
- — n x OC12/STM-4
- — n x OC48

1.1 INTRODUCTION

Technology has for many years now been governed by the law of exponentials. Hyper growth has occurred in microprocessor speeds, storage, and optical capacity. What have historically been independent developments have now come together to form a powerful technology ensemble on which new technologies such as grid computing can flourish. This chapter highlights the significant developments that have occurred in the past decade.

1.2 MICROPROCESSOR TECHNOLOGY

In a 1965 article in *Electronics* magazine, Intel® founder-to-be Gordon Moore predicted that circuit densities in chips would double every 12–18 months.[1] His proclamation in the article, "Cramming More Components onto Integrated Circuits," has come to be known as Moore's Law. Although Moore's original statement was in reference to circuit densities of semiconductors, it has more recently come to describe the processing power of microprocessors. Over the past 37 years, Moore's Law has successfully predicted the exponential growth of component densities, as seen in Table 1.1.

TABLE 1.1 Exponential Growth of Transistors on a Single Chip

Year	Transistors
1970	2,300
1975	6,000
1980	29,000
1985	275,000
1990	5,500,000
2000	42,000,000

The exponential growth of computing power is evidenced daily through the use of our personal computers. The personal desktop computer of today processes one billion floating point operations per second, giving it more computing power than some of the world's largest supercomputers circa 1990. The high-performance computing systems (a.k.a. supercomputers) have also seen many folds of growth.

---

[1] *Electronics*, Volume 38, Number 8, April 19, 1965.

For example, the peak performance of the fastest supercomputer in 1993 was 57.9 Gflops as compared to 40 TFlops in June of 2003.[2]

Today, Intel believes that it is on target to deliver a one billion transistor microprocessor by 2010. However, as densities and processing power of microprocessors have increased, so have their manufacturing costs. This phenomenon may well come to be known as Moore's Second Law, which he explained in an interview with *Wired* as:

> *The cost of manufacturing facilities doubles every generation. In the late 1980s, billion-dollar plants seemed like something a long way in the future. They seemed almost inconceivable. But now, Intel has two plants that will cost more than $2.5 billion. If we double it for a couple of generations, we're looking at $10 billion plants. I don't think there's any industry in the world that builds $10 billion plants, although oil refineries probably come close.[3]*

In 1995, the cost for a wafer fabrication plant was US $1 billion and approximately 1 percent of the annual microprocessor market. By 2011, the cost will increase to US $70 billion per plant and approximately 13 percent of the annual microprocessor market.

It is expected that optical lithiography—the main technology in wafer development—will bump up against the laws of physics in 2017. In the meanwhile, researchers are busy working on new technologies such as Nuclear Magnetic Resonance, Ion Traps, Quantum Dots, Josephson Junctions, and optical chips that will take computing power on its next exponential journey.[4]

1.3 OPTICAL NETWORKING TECHNOLOGY

Today, we can carry more traffic in a second, on a single strand of fiber, than all of the traffic on the whole Internet in a month in 1997.[5] This phenomenon has been made possible by developments in fiber optics and related technologies. These developments enable us to put more and more data onto optical fiber. In the mid 1990s, George Gilder predicted that the optical capacity of a single fiber would treble every 12 months. Therefore, it is no surprise that Gilder's Law is now used to refer to the exponential improvements in optical capacity.

---

[2] Muer, Hans, 2002, *Top500.org – www.top500.org.*

[3] Leyden, Peter, 1997, "Moore's Law Repealed, Sort of," in *Wired.*

[4] Porobic, Adi, 2002, *What Next, Revealed 1, www.geocities.com/adiporobic/techie.htm*

[5] World Economic Forum, New York 2001, *Digital Divide Report.*

One of the key optical technologies is called dense wave division multiplexing or DWDM. DWDM increases the capacity of the fiber by efficiently utilizing the optical spectrum. Certain frequency bands within the optical spectrum have better attenuation characteristics and have been identified as suitable for optical data communications.

The increase in optical capacity of a fiber occurs on two dimensions. First, the number of wavelengths utilized within the spectrum is increased and, second, the amount of data pushed through each wavelength is increased.

In the early 1980s, systems supported two widely separated wavelengths: one in 1310 nm and another in 1550 nm bands as shown in Figure 1.1. Each wavelength supported a few megabits per second. The next generation of systems that followed allowed 2–8 wavelengths on fiber. Over the years, the spacing between the wavelengths has decreased from 400 GHz to 12.5 GHz while the bit rate of each wavelength has increased from a few Mbps to 10 Gbps. A typical fiber system can now carry up to 640 Gbps.[6]

---

[6] *Corestream Product Specification*, Ciena Corporation.

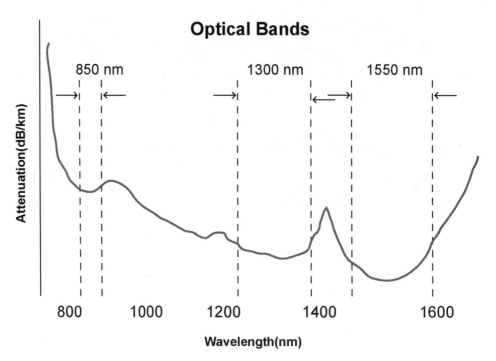

FIGURE 1.1   Optical networking spectrum.

Key developments in amplification technologies allow deployments of optical systems that run thousands of kilometers without any regeneration. All optical switches based on MEMS (Micro Electro Mechanical Systems) and other technologies allow switching to occur in the optical domain, thereby taking out costly electrical interfaces.

The advancement in optical technologies, telecommunication deregulation, and general euphoria around the dot-com economy led to a boom in construction of optical fibers all around the world. Thousands of miles of terrestrial and submarine cables were laid in the anticipation of demand that, to date, has not materialized.

Today only 10 percent of potential wavelengths on 10 percent of available fiber pairs is actually lit as shown in Figure 1.2. This represents 1–2 percent of potential bandwidth that is actually available in the fiber system. However minor this portion might seem, the amount of lit capacity it represents is staggering. For instance, in New York City alone, 23.5 Tbps run through the city on domestic and international networks.

The result of this severe imbalance between supply and demand has understandably led to tremendous price erosion of bandwidth products. Annual STM-1

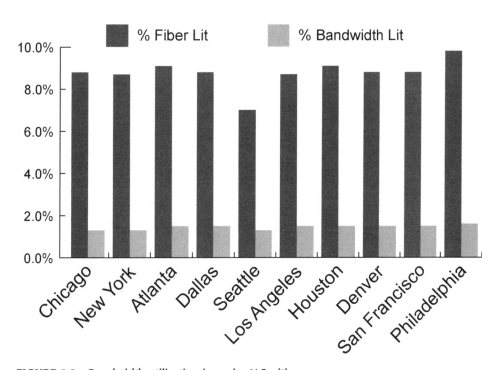

FIGURE 1.2 Bandwidth utilization in major U.S. cities.

(OC-3, 155 Mpbs) prices on major European routes have fallen by 85–90 percent from 1990 to 2002.

During the first quarter of 2000, an OC-3 (155 Mbps) lease between Los Angeles and New York went for US $18 million and two years later, the same lease traded at under US $190,000. At the time this chapter was written, the prices continued to spiral aggressively downward.[7]

1.4 STORAGE TECHNOLOGY

Fujitsu Corporation® recently announced that it had developed a new read/write disk head technology that will enable hard disk drive recording densities of up to 300 gigabits per square inch. Current 2.5-inch hard disks can store around 30GB per disk platter. When Fujitsu commercializes its new disk head technology in two to four years' time, it will lead to capacities of 180GB per platter—six times current capacities.[8] This growth in storage capacity is, in fact, not new. Over the last decade,

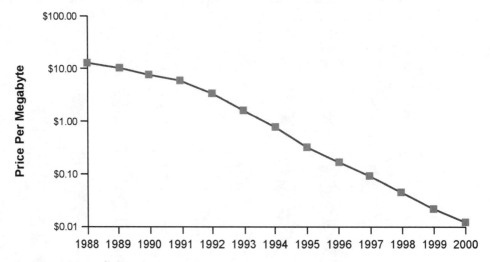

FIGURE 1.3   Declining storage costs.

disk storage capacity has improved faster than Moore's Law of processing power. Al Shugart, founder and former CEO of Seagate™, is sometimes credited with pre-

---

[7] Telegeography, Inc., *Terrestrial Bandwidth*, 2002.

[8] Fujitsu Computer Products of America Press Release, May 2002, "Fujitsu Develops Record-Breaking 300Gb/in_ HDD Technology."

dicting this; hence, for the sake of consistency, we will refer to the exponential increase in storage density as Shugart's Law.

But storage capacity is not the only issue. Indeed, the rate with which data can be accessed is becoming an important factor that may also determine the useful life span of magnetic disk-drive technology. Although the capacity of hard-disk drives is surging by 130 percent annually, access rates are increasing by a comparatively tame 40 percent.

To enhance access rates, manufacturers have been working to increase the rotational speed of drives. But as a disk spins more quickly, air turbulence and vibration can cause mis-registration of the tracks—a problem that could be corrected by the addition of a secondary actuator for every head. Other possible enhancements include the use of fluid bearings in the motor to replace steel and ceramic ball bearings, which wear and emit noticeably audible noise when platters spin at speeds greater than 10,000 revolutions per minute.

There will be a bifurcation in the marketplace, with some disk drives optimized for capacity and others for speed. The former might be used for mass storage, such as backing up a company's historical files. The latter would be necessary for applications such as customer service, in which the fast retrieval of data is crucial.

As storage densities have improved, their price has come down substantially. The price elasticity from storage remains greater than that of bandwidth or microprocessors. For every 1 percent decrease in price there is a 4 percent increase in usage.[9] However, manufacturers also have found it hard to turn their technical expertise into profit. As price of hard disk storage has fallen from US $11.54 per megabyte in 1988 to US $.01 per megabyte today (see Figure 1.3), the number of independent manufacturers has dropped from 75 to 13.

The demand for storage, however, is insatiable and will remain so for a long while. For example, 80 billion digital photos that would take more than 400 petabytes to store are taken each year.[10] Also, the demand for storage is not only on the rise in traditional segments such as personal computers and servers, but in other areas such as laser printers, GPS systems, set top boxes, digital cameras, and the gaming industry.

1.5 WIRELESS TECHNOLOGY

It is interesting to note that no luminary has been granted a law of exponential growth in his or her name in the field of wireless technology. Perhaps it is because

[9] Bruno, Lee, 2000, "Disk Driven Technology," *Red Herring*.

[10] University of California, Berkeley, 2000.

wireless technology has not seen the exponential gains in raw capacity that other technologies have. Nonetheless, the adoption rate of mobile wireless in particular has been no less than spectacular around the world. Many countries have started deployment of 3G technology that promises to bring data at a rate of 1 Mbps to a wireless device. The deployment of 3G now has more to do with the financial condition of the carriers, and whether they can actually afford to deploy these systems, than with the robustness of the underlying technology.

Fixed wireless, another technology that was supposed to revolutionize bandwidth access, has not fared that well. The adoption rate of licensed spectrum fixed wireless technologies such as MMDS (Multipoint Microwave Distribution System) and LMDS (Local Multipoint Distribution System) has been extremely disappointing. In the United States, AT&T™ announced that it will not be deploying an MMDS network, and another major carrier—Sprint™—recently announced that it would stop all MMDS deployment until next generation systems are available, sometime in 2005. Worldcom, one of the largest holders of MMDS licenses in the United States, has them on the auction block for reasons that can be determined from the headlines of any popular newspaper.[11]

However, there is a bright star in the wireless world. If early numbers are any indication, then deployment of wireless LAN networks is expected to be widespread and quite rapid.

Wireless LAN is governed by the 802.11 standards being developed at the IEEE. This is actually a family of standards:[12]

- 802.11—applies to wireless LANs and provides 1 or 2 Mbps transmission in the 2.4 GHz band using either frequency hopping spread spectrum (FHSS) or direct sequence spread spectrum (DSSS).
- 802.11a—an extension to 802.11 that applies to wireless LANs and provides up to 54 Mbps in the 5GHz band. 802.11a uses an orthogonal frequency division multiplexing encoding scheme rather than FHSS or DSSS.
- 802.11b (also referred to as 802.11 High Rate or Wi-Fi)—an extension to 802.11 that applies to wireless LANs and provides 11 Mbps transmission (with a fallback to 5.5, 2 and 1 Mbps) in the 2.4 GHz band. 802.11b uses only DSSS. 802.11b was a 1999 ratification to the original 802.11 standard, allowing wireless functionality to be comparable to Ethernet.
- 802.11g—applies to wireless LANs and provides 20+ Mbps in the 2.4 GHz band.

---

[11] On 6/25/2002 Worldcom declared that it had committed fraud by improperly capitalizing operating expenses, which resulted in a profit restatement of US $3.4 billion in 2001. Worldcom since then has decided to get out of its non-core business and has put its MMDS licenses on the auction block.

[12] IEEE 802.11x standards body.

The most widely deployed standard is 802.11b, or Wi-Fi, which allows transmission of wireless data at speeds as fast as 11Mbps and has a range of 300 feet to 800 feet. Boosters such as sophisticated antennas can be added to improve the range of the wireless network.

The cost of deploying a wireless LAN has come down substantially. There are now hundreds of thousands of "hot spots"—or Wi-Fi-enabled locations—around the world in airports, cafés, and other public places that provide unprecedented roaming capabilities.[13]

The benefits of 802.11b and 802.11a are not limited to mobile users or home applications, but will be some of the key technologies that reshapes the manufacturing and production operations in much the same way as the electric motor led the transformation of the sector in the early part of the 20th century. They will play an extremely important role in seamlessly deploying wireless networks at manufacturing plants, factory floors, and warehouses, thus providing a real-time view of operations, production, and inventory. We will expand on this line of thinking in the next section.

1.6 SENSOR TECHNOLOGY

Sensor technologies in one form or another have been with us for a while. However, recent advancement in certain types of technologies have brought their price and performance capabilities within reach for pervasive deployment. Paul Saffo of the Institute of the Future goes so far as to say that if the 1980s was the decade of the microprocessor and the 1990s the decade of the laser, then this is the decade of the sensor.[14]

A key advancement in sensors has been the ability to not only collect information about their surroundings, but also to pass the information along to upstream systems for analysis and processing through various integrated communication channels. In this section, we specifically focus on two such types of sensors.

1.6.1 Fiber Optic Sensors

Optical fiber sensors have been around since the 1960s when the Fotonic Probe made its debut. But the technology and applications of optical fibers have

---

[13] 802.11planet.com – INT Media's site for 802.11x news and information.

[14] Saffo, Paul, 2002, "Smart Sensors Focus on Future," *CIO Insight*. Institute of the Future is based in Palo Alto, California.

progressed very rapidly in recent years. Optical fiber, being a physical medium, is subjected to perturbation of one kind or another at all times. It, therefore, experiences geometrical (size, shape) and optical (refractive index, mode conversion) changes to a larger or lesser extent depending upon the nature and the magnitude of the perturbation. In communication applications, one tries to minimize such effects so that signal transmission and reception is reliable. In fiber optic sensing, on the other hand, the response to external influence is deliberately enhanced so that the resulting change in optical radiation can be used as a measure of the external perturbation. In communication, the signal passing through a fiber is already modulated, while in sensing the fiber acts as a modulator. Fiber also serves as a transducer and converts measurands like temperature, stress, strain, rotation, or electric and magnetic currents into a corresponding change in the optical radiation.[15]

Because light is characterized by amplitude (intensity), phase, frequency, and polarization, any one or more of these parameters may undergo a change. The usefulness of the fiber optic sensor, therefore, depends upon the magnitude of this change and ability to measure and quantify the same reliably and accurately.

The advantages of fiber optic sensors include freedom from EMI, wide bandwidth, compactness, geometric versatility, and economy. Absence of EMI and geometric versatility make them great candidates for biomedical applications.

Many of the new "smart" structures, such as bridges being built today, have an array of fiber optic sensors deployed to check for corrosion and other flaws that can lead to catastrophic failures. Fiber optic sensing, for example, is being used to detect damaged rails and faulty wheels on trains by firmly attaching ultra sensitive optical fibers to the rails. An environmental change, such as the weight of a passing train or the strain created by a cracked, broken, or buckled rail, is immediately detected and analyzed.[16]

1.6.2 Wireless Sensors

Another set of sensors are being developed with embedded RFID (Radio Frequency Identification Devices) tags. A consortium of companies under the aegis of the Auto-ID Center, have developed technologies that will in the very near future replace bar codes. These product codes will be embedded in a RF-enabled "smart tag" that will be attached to products and automatically read by scanners.[17]

---

[15] Culshaw, S., "Revisiting Optical Fiber-Sensors," *Optical Fiber Sensors*, Volume III, January 1997.

[16] Kloeppel, James E. 2001, *Fiber Optics Sensors Detect Damaged Rails and Faulty Wheels*, UIUC.

[17] Auto-ID Center is a global industry funded research program that was established at MIT in 1999 and Cambridge University in 2000. Additional labs are being set up in Asia and South America. Industry sponsors include Proctor & Gamble, Gillette, Johnson & Johnson, Wal-Mart, NCR, Sun Microsystems, Invensys, and Alien Technology.

A 96-bit code of numbers, called an Electronic Product Code (ePC), is embedded in a memory chip (smart tag) on individual products. Each smart tag is scanned by a wireless radio frequency "reader" that transmits the product's embedded identity code to the Internet or intranet where the "real" information on the product is kept.

Some features of this technology have already been implemented in the manufacturing and retail industries. For example, all clothes sold at Prada's flagship store in downtown New York City have these wireless smart tags. As soon as one of the items is brought into the dressing room, the LCD monitor there displays detailed information about the product to the customer. Meanwhile, Prada is collecting invaluable and realtime information about customer behavior and preferences.[18]

Additionally, there are sensors that integrate with RFID tags and provide information about the condition of products, in addition to their location, as they move through the supply chain. Such sensors can detect various types of damaging events, including impact, tilt, rollover, and load shifting. In the distribution of products from manufacturers to consumers, damage, due to improper packing and handling, totals more than $4 billion in unsaleable products per year. Studies show that nearly 50 percent of unsaleable goods resulted from products being crushed, dented, or collapsed.[19]

The ultimate goal is to have everything connected in a dynamic, automated supply chain that joins businesses and consumers together in a mutually beneficial relationship.

1.7 GLOBAL INTERNET INFRASTRUCTURE

The Internet started in the late 1960s when the United States Defense Department's Advanced Research Projects Agency developed ARPANET technology to link up networks of computers and, thus, make it possible to share the information they contained. During the 1980s, the National Science Foundation invested about $10 million per year, which motivated others (universities, states, private sector) to spend as much as $100 million per year to develop and support a dramatic expansion of the original ARPANET. The National Science Foundation Network (NSFNET) that resulted gave U.S. universities a nationwide, high-speed communications backbone for exchanging information. The network initially served academic researchers, nonprofit organizations, and government agencies. To meet a new demand for commercial Internet services, NSF, in 1991 decided to allow for-

---

[18] Personal visit by the author in 2002.

[19] Signal Quest Web site, *www.signalquest.com*

profit traffic over the network. During the next four years, NSFNET was privatized and became the global Internet we know today.[20]

The privatization of the Internet in the United States and the frenzy around the World Wide Web, e-commerce, and the like coincided with another key global event: the beginning of the deregulation of the telecommunications companies. In the United States, this started with the Telecommunications Act of 1996. Led by the World Trade Organization, deregulation around the world started shortly thereafter with 90 percent of the global telecommunications deregulation to be completed by 2004 and 99 percent by 2010.[21]

The privatization of the Internet and the deregulation of the telecom industry collided with a force that led to some of the most spectacular infrastructure investments of the century. The capital markets and the venture capitalists poured money heavily into Internet "dot-com" companies. The markets, with equal vigor, funded companies that were building the networks, the data centers, and the equipment that would serve the growing bandwidth demands of the "Internet-based" economy. Unfortunately, the demand for bandwidth and related services never materialized as the Internet economy ran into some old economic realities, such as the importance of free cash flow and profits.

The Internet boom lasted five years and collapsed in March of 2000. The telecommunication bubble, which started in 1996, today leaves in its wake more than 200 bankruptcies and liquidations. But it was exciting while it lasted.

Although the impact of the dramatic meltdown of the Internet and telecommunications segments cannot be understated, there is clearly a silver lining in the debacle. There are three developments that occurred during the "boom" period that we think are extremely encouraging.

The first, and the most significant, change that has occurred over the last few years is a deep appreciation for open and standards-based systems. Many of the companies that were adherent to proprietary, closed end systems were forced to take a close look at open systems and by the end of the 1990s most were converted. There are dozens of standards bodies—such as the Optical Internetworking Forum, Metro Ethernet Forum, 10 Gigabit Ethernet Alliance, W3C, Internet Engineering Task Force (IETF)—that were formed in this "bubble" period and are still around today and doing excellent work in delivering standards. The acceptance of open, standard-based systems will have far reaching benefits in the years to come.

---

[20] National Science Foundation Web site, *www.nsf.gov*

[21] Hudson, Heather E., 1997, "Privatization and Liberalization in the Developing World: The Need for Innovative Policies and Strategies," ITU, 1997. *World Telecommunication Development Report 1996/97* "Trade in Telecommunications" (Executive Summary). Geneva, Switzerland: International Telecommunications Union.

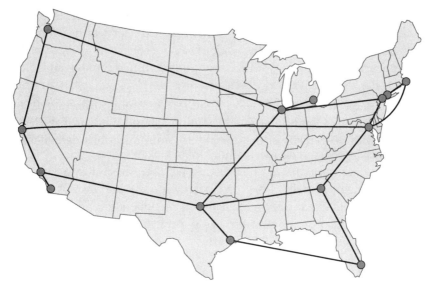

FIGURE 1.4   UUNET network in January of 1996.

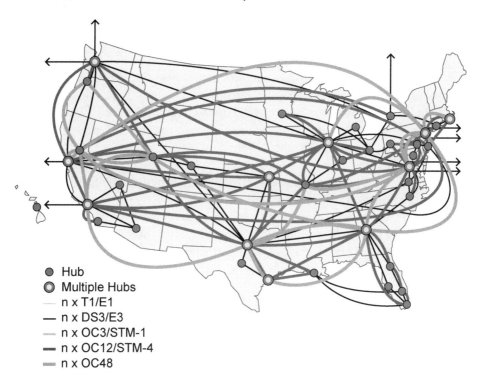

FIGURE 1.5   UUNET network in September of 1999.

The second key benefit is the substantial research, development, and implementation of security, security standards, and encryption for the public Internet. Although there is still a lot of work to be done, over the past few years the public Internet has become a powerful conduit for commerce and business.

The third important development is the deployment of an impressive global networked infrastructure that is now ready to be harvested by new technologies, such as Grid Computing. The Internet was successfully transformed from a playground for academics to one that can withstand the scale and complexity of commercial deployment. In addition to the tremendous optical fiber deployment around the world discussed in section 1.2, there are now more than 2000 data centers available globally, 140 global peering points, and broadband deployment initiatives in all Organization for Economic Cooperation and Development (OECD) countries.[22] Korea, for example, has 85 percent broadband penetration while China will be deploying 1,000,000 Ethernet connections directly to the home in its 10th Five Year Plan 2001–2005.[23]

Technologies, such as grid computing, that rely on a networked infrastructure, will no doubt take advantage of the deployed infrastructure.

1.8 WORLD WIDE WEB AND WEB SERVICES

The Internet was 23 years old when in 1993, the World Wide Web (through HTTP/Mosaic browser)[24] was introduced to the public. The World Wide Web, much more so than e-mail, turned out to be the "killer app" that led to the expansive growth of the Internet. Today there are roughly 200 million hosts on the Internet, up from 1 million in 1993.

The Internet and the World Wide Web have had a substantial impact on the lives of peoples throughout the world. Many credit the Internet for the business productivity miracle in the U.S. economy that started in 1995—the same year that Netscape went public. It is estimated that U.S. companies have saved US $251–400 billion in the past five years by using the Internet. It has also been determined that much of the savings comes from bringing mundane transactions that involve information flow, ordering, invoicing, filing claims, etc., online.[25] This number represents 1–1.5 percent

---

[22] Telegeography, "Colocation 2002."

[23] Abbas, Ahmar, *Optical Networking in China*, Presentation at China Telecom 2002, Honolulu, HI.

[24] HTTP was developed at CERN and Mosaic browser at NCSA—both organizations today are at the forefront of Grid Computing activities.

[25] Litan, Robert E. and Rivlin, Alice M. 2001, *Economic Payoff from the Internet Revolution*, Brookings Institution Press.

of the total U.S. economy. But we believe, as do many others, that these figures are really the tip of the iceberg and only represent savings derived by a very small segment (the computer industry itself) of companies.[26] We classify most of the activities on the Web today to be consumer focused. In other words, we are living in the age of the "consumer Web." However, a set of standards widely embraced by the software and hardware vendors are setting the foundation for the "business Web," where the value of the Internet can be truly brought to bear to increase corporate productivity. These standards are collectively called "Web Services."

Web Services promise to do away with the ad hoc interaction among businesses by providing a systematic and extensible framework for application to application interaction, built on existing Web protocols and based on open standards. Unlike traditional client/server models such as the Web server/Web page system, Web Services do not provide the user with a graphical user interface. Web Services, instead, share business logic, data, and processes through a programmatic interface across a network. Web Services allow different applications to communicate with each other without time consuming custom coding. Web Services, being XML-based, are not tied to any one operating system or programming language. For example, Java™ can now talk to Perl™ and Windows® applications can now talk to Unix™ applications.

The Web Services framework is divided into three areas—communication protocols, service descriptions, and service discovery—with specifications being developed for each. The specifications that are currently the most salient and stable in each area are as follows:[27]

■ SOAP—Simple Object Access Protocol enables communication among Web Services.
■ WSDL—Web Services description language provides a formal, computer-readable description of Web Services.
■ UDDI—universal description, discovery, and integration directory serves as a registry of Web Services descriptions.

To understand how Web Services work, let us examine their application to the travel industry. The global travel and tourism industry generates US $4.7 trillion in revenues annually and employs 207 million people directly and indirectly. It also drives 30 percent of all e-commerce transactions.[28] However, one of the dominant

---

[26] Grid Technology Partners, 2002, *Computer Paradox and Grid Computing, www.gridpartners.com/articles.html*

[27] Web Service Group, World Wide Web Consortium, W3C—*http://www.w3.org/2002/ws/*

[28] Pollock, Anna & Benjamin, Leon, 2002, *Why Web Services Will Turn the Travel Industry on Its Head—and Why that's a Good Thing!*

features of travel and tourism is supply side fragmentation, extreme heterogeneity and diversity in terms of focus and size of the travel service providers. On the demand side, there is a need to cater to a consumer's total trip and deliver a total experience. Web Services would allow each provider to advertise their services (just as they may do today on their Web sites) in a machine readable format. XML would be used to describe the data and WDSL would specify the Web Service. UDDI would be used to list the service in a directory and the SOAP protocol would be used for communication on the Internet. An online travel agency would be able to perform searches for various services and book the services for the customer. The same agent would be able to connect to the customer's bank—again using Web Services—to order traveler's checks. The agent would also be able to put customer's newspaper subscription and mail on hold through Web Services-based interactions with the newspaper and the post office.

Web services are being championed by Microsoft®, SUN®, IBM®, and a host of other key players in the market space. Microsoft .NET initiative™ and Sun Microsystems' SUN One™ are two of the more popular Web Services initiatives.

1.9 OPEN-SOURCE MOVEMENT

The ubiquity of the Internet and the World Wide Web has also spawned a powerful movement for open-source software development. The open-source movement is helping turn significant chunks of the IT infrastructure into commodities by offering free alternatives to proprietary software. The promise of the past several years has begun to materialize as, one by one, the hurdles to open-source adoption have dropped away. Major enterprises are running mission-critical functions on open-source IT. Big vendors have lined up to support it or port their applications to it. CIOs (chief information officers) who have implemented it report significant reductions in total cost of ownership. Starting in 2001 and 2002, major vendors such as Dell™, HP™, IBM, Oracle™, and Sun announced in various ways that they would begin supporting open-source products.

1.10 CONCLUSION

Advances in microprocessor, networking, and storage technologies have been nothing short of astounding. This core technology ensemble has contributed to creating an extremely powerful infrastructure for consumer and business applications. The timing is right for new technologies, such as grid computing, to leverage the ever-increasing power and declining costs of these core technologies.

The power of the desktop has become too significant to ignore. The exponential increase in microprocessor and storage capacity has made desktop computers extremely powerful resources. The near ubiquitous deployment of these computers provides companies with latent resources that can be utilized by grid technologies to make a substantial impact on business processes and increase productivity.

The global telecommunication infrastructure built over the past few years is mostly unused today. The price of telecommunication services have also dramatically come down, thus making it easier to build and acquire high-speed networks for Grid Computing applications.

We have developed an appreciation for standards. Many of the companies that were adherent to proprietary, closed end systems were forced to take a close look at open systems and, by the end of the 1990s, most had converted. This is already benefiting the Grid Computing technology evolution. Vendors, large and small, have committed from the beginning to open standards for Grid Computing and are active participants in their definition.

2 Productivity Paradox and Information Technology

Ahmar Abbas

Grid Technology Partners

In This Chapter

- Information Technology (IT) Failures and Challenges
- Infrastructure Consolidation
- Real-Time Enterprise

© Grid Technology Partners

2.1 INTRODUCTION

Information Economy and *New Economy* are some of the terms that have been used to describe the technology driven economies of most of the industrialized countries. As technology has proliferated in enterprises, its anticipated effect on corporate productivity, however, has not been that apparent. This chapter discusses that state of information technology and some of the challenges being faced by enterprises in dealing with IT.

2.2 PRODUCTIVITY PARADOX

It is a little known fact—one that goes against all that we, as technology acolytes, have held dear to us—that the productivity growth rate in the United States attributed to technology (also known as the Total Factor Productivity or TFP), as defined by the keen economists at the Federal Reserve Bank, slowed down sharply, from 1.9 percent between 1948 and 1973 to only 0.2 percent from 1973 to 1997. This period completely overlaps with the introduction and the amazing proliferation of desktop and other computing resources in U.S. companies and homes. This became known as the "computer paradox" after MIT professor and Nobel laureate Robert Solow made the famous 1987 quip: "We see the computer age everywhere except in the productivity statistics."

The statistics for the last few years are a bit more heartening. Productivity growth has averaged considerably higher at 2.7 percent per annum. Some say that this is due to the "diffusion effect"—that we are just now reaping the benefits of our earlier information technology investments. However, others argue that it is actually due to the "concentration effect" and that the entire acceleration of productivity growth can be traced directly to the computer manufacturing industry itself. This one industry, while just a small piece of the overall economy, has produced a genuine productivity miracle since 1995.

2.3 RETURN ON TECHNOLOGY INVESTMENT

While economists and researchers continue to study the impact of technology on productivity and corporate profits, corporations have drastically modified their thinking on technology spending. It is no longer sufficient to base technology purchase decisions solely on the technological merits or glamorous features of products. A recent study of Siebel™ customers revealed that 60 percent of them did not believe that they received any significant return on their investment in various customer relationship management (CRM) software programs that they purchased from Siebel. Standish Group's recent study indicates that of the US $250 billion spent annually on technology projects, 30 percent partially fail, 52 percent are severely challenged, and 16 percent completely fail, and another study referenced by Verizon™ indicates an 80 percent technology failure rate. In simpler terms, the romance with technology is over and laws of economics have taken over.

During the last decade or so, the decision to purchase and deploy technology was largely made by the chief information officer (CIO) and driven by the business unit. The chief financial officer (CFO) was brought into the discussion to negotiate the best deal possible from the vendor. However, in recent times, the chief financial officer is becoming more involved and usually requiring detailed analyses of financial measures such as net present value (NPV), return on investment (ROI), internal rate of return (IRR), and payback period.

It is therefore not a surprise that today's software and hardware vendors, while highlighting product features, are also focusing a lot of effort on presenting the financial benefits of their products.

2.4 MULTI-STORY BUREAUCRACY

Failure of information technology to increase productivity and have an impact on corporate profitability is in itself not unprecedented. One can find an interesting parallel in the adoption of electricity in the early 20th century. As the 20th century began, business investment in electrical equipment skyrocketed. Between 1900 and 1920, the percentage of U.S. factories equipped with electric motors jumped from 5 percent to 55 percent. Yet, despite the obvious advantages of electric motors over steam engines, worker productivity showed almost no measurable increase.

Solving the riddle of today's computer productivity paradox is simple, once its century-old predecessor is understood. Back then, a state-of-the-art facility was a three- or four-story brick building. A coal-fired steam engine sat in the factory's basement. Its power was transmitted to the equipment on the floors above through an elaborate system of vertical and horizontal shafts and drive belts.

Factory owners began replacing steam engines with large electric motors because the new technology slashed coal bills by 20 to 60 percent. But the power generated by these first electric motors was still conveyed to lathes, drills, grinders, and punch presses by the conventional system of shafts and belts. As time passed and somewhat smaller electric motors became affordable, separate motors were installed on each floor. This eliminated the need for the vertical shaft. However, because the motors were still relatively expensive, each one powered a group of machines via horizontal shafts and belts.

Electric motor technology was indeed revolutionary, but to pay for itself, it had to be grafted onto the existing industrial infrastructure. Multi-story factories made sense in the steam age because they reduced the costly friction losses incurred as horizontal shafts carried power to the equipment. Single-floor factories would have eliminated all the labor wasted moving unfinished goods from floor to floor, but overall it was cheaper to staff the elevators than to pay for the coal turned to waste heat by long horizontal shafts.

For similar reasons, machines requiring the greatest amount of horsepower were placed near the base of each floor's horizontal shaft. Smaller, low-power-consuming equipment sat at the far end of the work floor. Again, although this arrangement made economic sense from a power-conservation standpoint, it made the flow of materials ridiculously inefficient. In short, because the entire industrial infrastructure had evolved around the assumption of expensive steam power, factories were designed to be grossly inefficient in other dimensions. Once made, these fundamental economic tradeoffs endured for decades in bricks and mortar.

It wasn't until cheap, small electric motors became available in the 1920s that factories began to abandon "group drive" power for the "unit drive" approach—the familiar present-day system in which each machine is powered by its own internal motor. As this technology was installed, companies ripped out their drive shafts and belts, rearranged their machines, and smoothed their flow of materials. The most innovative "high-tech" firms—companies in fast-growing new industries such as cigarettes, organic chemicals, and electrical equipment—were the first to go all the way and take the radical step of building single-story factories. They were literally the first "flattened" corporations.

To reap the full promise of electricity, an almost endless number of practical details had to be worked out. That took time and enormous creativity. Finally, 40 years after Edison's breakthroughs, productivity growth took off. Where the annual labor productivity growth had hovered around one percent for the first two decades of the century, it jumped to more than five percent a year during the 1920s.

2.5 INFORMATION TECHNOLOGY STRAIGHTJACKET

We live in "multi-story" technology and organization bureaucracies that are in great need of being flattened out into "single-story" efficient organizations. The change has begun—not because of any sudden shift of heart by top management but because the hard economics of information handling are being transformed. At stake today is the competitive survival of the corporation.

In the electricity age, no attention was paid to the business processes in creating the manufacturing environments. The same can be said of technology infrastructure deployment, which tend to be mostly *point* solutions rather than being an integral part of the business. There has been a perennial lack of progress in areas such as:

- Integration—systems, process, data
- Data Hygiene—consistency, integrity, security
- Data Models—lack of semantic understanding

Information technology deployments today are marked with batch or manual processes. Many systems within the same enterprise are disconnected and exist in silos. There is limited "partner" connectivity and, as a whole, corporate IT infrastructure contains a tremendous number of low value human touch points. In many cases, the value of the information loses its currency as it is manually processed.

The reality today is that information technology is a constellation of system clusters. Each cluster contains many applications, databases, servers, networking infrastructure, and its own management domain, as shown in Figure 2.1. Not only does the variety of applications and hardware continue to grow, but so does the scale and number of these deployments.

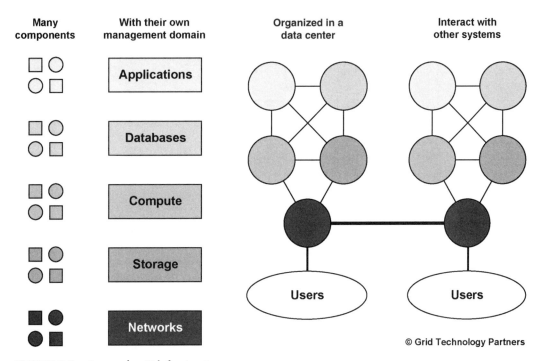

© Grid Technology Partners

FIGURE 2.1   A complex IT infrastructure.

Labor costs are escalating and in most cases they are the dominant cost (i.e., more than capital cost). In the storage segment, for example, the labor cost is approaching 3X the hardware cost.[1] According to various industry estimates, indirect costs may contribute up to 60 percent of overall total cost of ownership.

The information technology straightjacket comes about as technology complexity increases, labor costs increase tremendously, and the dependence on information managed by this technology also increases. Maintenance spending on technology overtakes any capital spending and—in many cases—any benefits accrued from the technology investment.

Companies are responding to this imbalance by focusing heavily on reduction of cost of their IT infrastructure. Figure 2.2 shows some of the traditional responses that we have seen from companies.

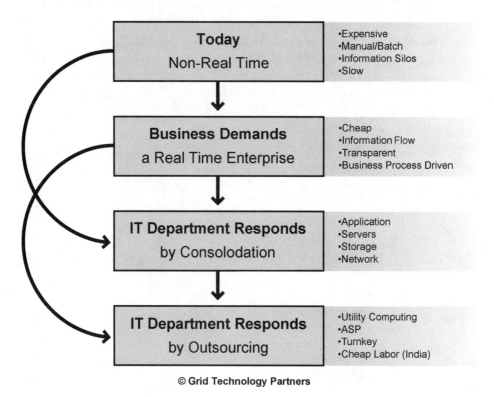

© Grid Technology Partners

FIGURE 2.2 Evaluating IT options.

---

[1] Based on US$ 120K loaded person cost, storage hardware at US$ 120K/TB with a 4-year life and IT Centrix survey result of 0.83 person-year/TB.

2.6 CONSOLIDATION

One of the first responses from organizations to the burgeoning IT costs and increased complexity has been to initiate "consolidation" projects. Initial emphasis has been primarily on consolidating computing resources, such as servers. However, recently the emphasis has expanded such that there are now corporate initiatives to consolidate storage resources, applications networks, data centers, process, and staff.

2.6.1 Server Consolidation

Servers are still the primary focal point for consolidation because they are obvious. Whether companies have 100 servers or 5,000, it is still a challenge to manage them effectively. Today's distributed computing environment lends itself to a proliferation of servers. Reducing and controlling the number of devices to manage and simplifying ways to manage them is the goal of most IT groups.

Servers can be consolidated *vertically* or *horizontally*. Vertical consolidation requires consolidating multiple applications onto a single server. Horizontal consolidation implies distribution of workloads across those servers.

However, not all servers are alike. One way to categorize servers is based on their functions.[2] The first function is the presentation tier. These are the servers that are closest to the user. Next is the business or middleware tier. This is where the applications or middleware run in conjunction with the other tiers. The resource tier is where the large, scalable servers run mission-critical applications and databases. Although consolidation (either vertical or horizontal) can occur at all of these tiers, it is important to make sure that there is no impact on the service level.

Consolidation can be physical, logical, or involve rationalization. Physical consolidation involves consolidating data centers and moving servers to fewer physical locations. The theory behind physical consolidation is that by having servers in fewer physical locations, you can achieve management consistencies and economies of scale more easily than you can when your servers are dispersed. Physical consolidations may also enable you to reduce real estate costs. It is generally felt that physical consolidation has the lowest risk, but that it also has the lowest payback.

Logical consolidation involves implementing standards and best practices across your server population. By doing this, you can realize substantial benefits in the productivity of your IT staff. They can manage the environment more efficiently and more effectively. This can often result in lower systems management

---

[2] The three-tier services model discussed here has been proposed by Sun Microsystems.

costs and in lower TCO. Logical consolidation is often implemented with physical consolidation and rationalization.

Rationalization involves the deployment of multiple applications on fewer, larger servers and in fewer instances of the operating system (OS). Because total cost of ownership (TCO) reduction is closely tied to the number of instances of an OS you manage, reducing the number is the best way to reduce TCO. Although rationalization is the riskiest form of server consolidation, it offers the biggest TCO reduction and return on investment (ROI).

2.6.2 Consolidating Applications

Understanding which server resources applications use and understanding usage patterns is essential to the process of consolidating servers. A recent survey revealed that 51 percent of the companies had more than 51 distinct applications as shown in Figure 2.3.

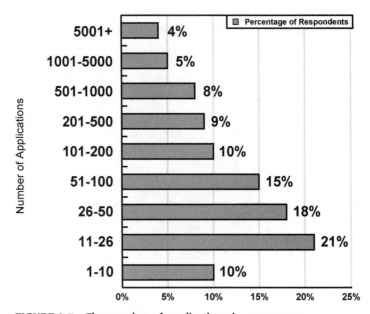

FIGURE 2.3 The number of applications in a company.

Because of the number of applications at enterprises, however, consolidation is usually not a trivial process. Tools such as MicroMeasure from Ejasent or BMC's Patrol and Perfect can help. However, a keen understanding of the application and its role in the enterprise is essential.

Application consolidation can occur in two phases. In the first phase, companies focus on existing applications. This process is known as *backward consolidation*. In many cases, these consolidations are hugely successful. In others, organizational issues prevent them from achieving their goals. There is no doubt that some of the largest and most successful consolidations we have seen have been backward consolidations.

To avoid future server sprawl, it is also important to begin to develop new applications in a consolidated environment, referred to as *forward consolidation*. It's interesting to note that applications that are developed in a consolidated environment usually behave with each other and can be moved into production in a consolidated environment.

2.6.3 Consolidating Storage

Today, there is as much interest in storage consolidation as there is in server consolidation. Every new server results in more storage. In many cases, the cost of storage exceeds the cost of the server and, although the server may not grow very much, the amount of storage required for an application will grow indefinitely.

Storage consolidation can occur through a variety of techniques. Apart from cost savings the additional benefits of storage consolidation include:

- Easier backup and recovery
- Increased availability
- Improved scalability
- Storage resource pooling

When undertaking a storage consolidation effort, it is also important to understand how disk space is utilized.

2.7 OUTSOURCING

Outsourcing IT operations can occur in many different ways. One way is through an *application service provider* model. In this case, the company only pays whenever it uses a particular application. The application itself resides outside the premises of the company and is accessed through the Internet or through some private network.

As applications have become more complex, many Independent Software Vendors (ISV) are realizing that the cost of maintaining and managing applications is deterring companies from purchasing them. In some cases, only the very large companies can afford their software. Fluent, the world's largest computation fluid dynamics software company has started to offer its products via the Application Service Provider (ASP) model. Their software had an initial cost of US $150,000 and a US $28,000 per seat licensing cost, but they are now offering their product for US $15.00 per CPU/hour. Now, not only do the customers not have any upfront cost, they have no hardware or maintenance costs. We see this model of outsourcing becoming increasing popular. Salesforce.com, which offers its sales force management software online, has successfully competed head-to-head with Oracle and Peoplesoft.

One variant of the ASP model is the new utility computing model of outsourcing. We will discuss this in more detail in the later chapters of this book. However, similar to the ASP model, which reduces the application costs on a pay-per-use basis, utility computing allows companies to pay only for the amount of computational, storage, or network capacity they use.

Some companies have been exploring outsourcing back-office to international IT companies in India, China, Costa Rica, etc., to save money. IT costs can be reduced by up to 40–50 percent in many cases.

Another method of outsourcing is the complete handing over of IT departments to systems integrators such as IBM's global services division. JP Morgan Chase recently outsourced its entire IT department to IBM.

2.8 TOWARD A REAL-TIME ENTERPRISE—OPERATIONAL EXCELLENCE

The latest buzz in information technology is the creation of the *real-time enterprise*. Real-time applications can help companies gain competitive advantage by being more responsive and proactive in the face of current business events. Unlocking real-time benefits will require improved integration both inside and outside the firewall. Robust workflow and process automation software will be needed to maximize the business benefit from the more timely data. One definition of real-time enterprise comes from uber-venture capitalist Vinod Khosla of Kleiner Perkins Caulfield Byers, who calls it: "Spontaneous Transaction Flow and Information Transparency throughout an Extended Enterprise Minimizing Latency and Labor."

In today's IT environment, there are multiple user interfaces that operate in silos. Predefined user processes are provided by packaged applications, while data are usually delivered in batch. Customization requirements usually require code modification by developers.

The real-time enterprise approach envisions user-centric processes, delivered via one portal framework. Workflow applications are configured by business analysts, rather than technologists. Process and data are gathered from disparate applications and powered in a heterogeneous environment.

The impact of the real-time corporation is that people will be able to make decisions quickly and accurately. Serial, manual processes give way to automated process management and activity monitoring. Operational cost and latency is reduced.

The underlying architecture for the real-time enterprise is dependent on successful development of policy-based systems that can adapt to the changing business environment. Grid Computing, as will be discussed throughout the book, addresses each of the challenges faced by the IT manager today.

2.9 CONCLUSION

The promise of information technology has yet to be fully realized. Although IT has proliferated at an astounding pace throughout enterprises, its impact on productivity has been muted as a result of ever-increasing complexity of deployments. The operational expenses associated with managing this complexity are eating into any benefits being derived from the technology. Enterprises need technologies that shield the complexity from users and IT staff and allow the infrastructure to be efficiently utilized. Grid Computing promises to be one such technology.

3 Business Value of Grid Computing

Ahmar Abbas

Grid Technology Partners

In This Chapter

- Building a Business Case for Grid Computing
- Grid Computing Marketplace

Grid Application / Service Provider			Grid Resource / Fabric Providers
Paramater Studies Data Mining Supply Chain	**Applications**	Collaboration Data Analysis Web Services	
Billing QoS Trading Portal Services	**Service Ware**	CRM Operations Management	
Agent Technology Directories Scheduling	**Middle Ware**	Programming Tools Security Resource Managers	
Network Resources Computers Storage	**Fabric**	Instruments Sensors Data Sources	

3.1 INTRODUCTION

Companies rely on information technology more so today than they ever have in the past. Trillions of dollars have been spent in the last decade by companies trying to optimize all aspects of their operations, such as financial, audit, supply chain, back office, sales and marketing, engineering, manufacturing, and product development.

Yet, enterprises today stand at a crossroads; most of the investments made in technology, as illustrated in Chapter 2, have yet to achieve their financial objectives or provide the expected boost in corporate productivity.[1]

Additionally, much of the deployed infrastructure remains hopelessly underutilized, as shown in Table 3.1.[2]

TABLE 3.1 Inefficient Technology Infrastructure

IT Resource	Average Daytime Utilization
Windows Servers	<5%
UNIX Servers	15 – 20%
Desktops	<5%

Grid Computing, a rapidly maturing technology, has been called the "silver bullet" that will address many of the financial and operational inefficiencies of today's information technology infrastructure.

After developing strong roots in the global academic and research communities over the last decade, Grid Computing has successfully entered the commercial world. It is accelerating product development, reducing infrastructure and operational costs, and leveraging existing technology investments and increasing corporate productivity. Today, Grid Computing offers the strongest low cost and high throughput solution that allows companies to optimize and leverage existing IT infrastructure and investments.

---

[1] Standish Group Survey, *The Chaos Report*. Project failure rate estimated US$ 250B/year. Thirty percent of projects failed, 52 percent are "challenged," and 16 percent succeeded; P. Strassman, *The Squandered Computer*, Information Economics Press, 1st edition, (April 1997); Grid Technology Partners, A. Abbas, *Computer Paradox and Grid Computing*, June 2002, GridToday.

[2] IBM Corporation, Taurus—Taxonomy of Actual Utilization of Real UNIX and Windows Servers—GM13-0191-00; Average daytime utilization across Windows servers is less than 5 percent, while corresponding aggregate daytime utilization of Unix servers is usually in the range of 15–20 percent. Daytime desktop utilization is usually less than 10 percent—and drops to being negligible in the evenings.

This chapter outlines key factors that should be considered by enterprises as they build business cases for deploying Grid Computing solutions. A *Values Analysis* is conducted and various *value elements* are discussed in detail. Additionally three types of risks are evaluated.

3.2 GRID COMPUTING BUSINESS VALUE ANALYSIS

In the post-bubble economy, value is more important than ever. In this section, we highlight some of the key value elements that provide credible inputs to the various valuation metrics that can be used by enterprises to build successful Grid Computing deployment business cases. Each of the value elements can be applied to one, two, or all of the valuation models that a company may consider using, such as return on investment (ROI), total cost of ownership (TCO), and return on assets (ROA).

3.2.1 Grid Computing Value Element #1: Leveraging Existing Hardware Investments and Resources

As outlined in Table 3.1, there is a tremendous amount of unused capacity in IT infrastructure at a typical enterprise. Grids can be deployed on an enterprise's existing infrastructure—including the multitude of desktops and existing servers—and thereby mitigate the need for investment in new hardware systems. For example, grids deployed on existing desktops and servers provide over 93 percent in up-front hardware cost savings when compared to High Performance Computing Systems (HPC).[3]

Costs savings are not limited to diminished hardware and software expenditure, but are also being derived by eliminating expenditure on air conditioning, electricity, and in many cases, development of new data centers. It is important that these savings not be overlooked when comparing Grid Computing solutions with other computing options.

Clusters, for example, are heavy pieces of equipment. A single cabinet containing just 16 nodes can weigh approximately 1200 lbs. It and its accessories will produce more than 20,000 BTUs an hour and will need to be kept between 60 F and 68 F.[4] The additional cooling requirements can easily overwhelm even the best of commercial cooling systems and may require expenditure on supplemental cooling

---

[3] Compared with IBM ASCI White. The savings are even larger when compared to multi-way systems from Sun and HP. Analysis by Grid Technology Partners in *The Global Grid Computing Report*, 2003.

[4] A rule of thumb is that electronics reliability is reduced by 50 percent for every increase of 18 degrees F above 70 F. Hence, insufficient cooling, even for a short period of time can be devastating to all electronics in the data center.

systems. Additionally, new power sources may be required if the existing ones cannot handle the two 15 amp circuits required for each 16 node system.[5] If the above environmental, power, and spatial requirements cannot be met within the existing data center, then a new one will have to be constructed (or leased). A current Tier IV data center build cost is US $6,600.00 per 16-node HPC system and it generally takes 15 to 20 months to construct.[6] These non-trivial infrastructure upgrade expenses usually alone are enough justification for leveraging existing computational resources for deploying grids.

3.2.2 Grid Computing Value Element #2: Reducing Operational Expenses

Grid Computing brings a level of automation and ease previously unseen in the enterprise IT environments. Key self-healing and self-optimizing capabilities free system administrators from mundane tasks and allow them to focus on high-value, important system administration activities that are longer term and more interesting. The ability of grids to cross departmental and geographical boundaries uniformly increases the level of computational capacity across the whole enterprise and enhances the level of redundancy in the infrastructure. This is a major breakthrough for system administrators who always seem to be chasing systems outages.

The operational expenses of a Grid Computing deployment are 73 percent less than for comparable HPC-based solutions.[7] Many of the existing cluster solutions are based on open-source and unsupported and complex cluster management software. Operational expenses associated with these deployments have been so high that many enterprises are being forced to outsource management of HPC systems to their suppliers.

Additionally, grids are being deployed in enterprises in as quickly as two days, with little or no disruption to operations. Cluster system deployments, on the other hand, are taking 60–90 days, in addition to the days required to configure and deploy the applications. Deployments may take longer if the existing enterprise data center is out of capacity. Additionally, HPC installations in data centers can cause substantial disruptions and potential downtime. In fact, 54 percent of data center site infrastructure failures are coincident with human activities.[8]

---

[5] Linux NetworX, *Planning for Your New Arrival*, 2001.

[6] W. Pitt Turner, *Industry Standard Tier Classifications Define Site Infrastructure Performance*, The Uptime Institute.

[7] Grid Technology Partners analysis based on 600 Gflop grid(desktop and server) and equivalent HPC solution.

[8] Kenneth Brill, *We Have Met the Enemy of Site Uptime and He is US!*, The Uptime Institute.

3.2.3 Grid Computing Value Element #3: Creating a Scalable and Flexible Enterprise IT Infrastructure

Traditionally, IT managers have been forced into making large-step function increases in spending to accommodate slight increases in infrastructure requirements. Even sparsely populated symmetric multi-processor (SMP) machines are quite a capacity and cost overshoot. IT managers either spend the money for a system that remains underutilized for some time or force the users to live under the tyranny of an overloaded system until such a time as the load justifies purchase of another system. Neither scenario is tenable in a fast evolving business environment.

Grid Computing allows companies to add resources linearly based on real-time business requirements. These resources can be derived from within the enterprise or from utility computing services.[9] Never again do projects have to be put on hold for a lack of computational capacity, data center space, or system priority. The entire compute infrastructure of the enterprise is available for harnessing. Grid Computing can help bring about the end of departmental silos and expose computational assets curtained by server huggers and bureaucracy. Yet, while departments will be making their resources accessible to the whole enterprise, Grid Computing still allows them to maintain local control.

3.2.4 Grid Computing Value Element #4: Accelerating Product Development, Improving Time to Market, and Raising Customer Satisfaction

Earlier, we focused on the key cost savings derived from grid deployments. However, in addition to cost savings, Grid Computing has a direct impact on the top line by accelerating product development at enterprises and helping bring product to market quicker. The dramatic reduction in, for example, simulation times can get products completed quickly. This also provides the capability to perform a lot more detailed and exhaustive product design, as the computational resources brought to bear by the grid can quickly churn through the complex models and scenarios to detect design flaws. Consider the following potential impact of Grid Computing solutions on the life sciences industry.

Life Sciences: Drug Discovery

Today, introducing a "New Chemical Entity," more commonly know as a "drug," into the market costs US $802M and takes an average of 12–15 years.[10] Almost

---

[9] Gateway Processing on Demand service offers CPU capacity at 15 cents a CPU/Hour.

[10] J.A. DiMasi, "The Price of Innovation: New Estimates of Drug Development Costs," *Journal of Health Economics* 22 (2003) 151–185.

80 percent of the NCEs actually never make it to market. The drug development costs are incurred in the pre-clinical as well as the clinical phases of drug development. Half of the cost in developing a drug is direct labor cost.

With such high stakes, life sciences companies, large and small, have turned to Grid Computing to shorten the drug discovery and development process. Grid Computing is being used in both the drug discovery phase to screen suitable, drug-like molecules against disease targets and also for clinical simulation, healthcare ecosystem modeling, and pharmacokinetic simulations. In short, Grid Computing is allowing drug companies to get the most out of their R&D expenditure by developing the *right* product and getting it to market in the shortest possible time. Companies can save almost US $5M per month in R&D expenses for each month shaved off the drug development process.[11]

The top line implications for the drug industry are as staggering and can amount to almost US $1M per day for each day that the product is brought to market early. [12] Additionally, reducing the time to market for drugs affords them longer patent protection and hence helps protect them from revenue cannibalization from generics.

In addition to life sciences, virtually all other sectors—financial, insurance, automotive, defense, etc.—are gaining competitive advantage by deploying grid solutions. Many of these sectors are discussed in later chapters.

3.2.5 Grid Computing Value Element #5: Increasing Productivity

Enterprises that have deployed Grid Computing are seeing tremendous productivity gains. Consider for example the productivity gains of an electronics design and automation company; run times of jobs submitted by its engineers were reduced by 58 percent by deploying a grid. Corporate-wide productivity gains by this reduction have been assessed at US $9M annually. [13] Such drastic reductions in run times and associated employee productivity gains are being seen in grid deployments in various enterprises.

Productivity gains serve as a crucial measure in any business analysis because they have a direct impact on the corporate bottom line. Yet in many instances of technology deployments, calculating productivity gains is more voodoo than science.[14] However, in the case of Grid Computing, enterprises have found it is

---

[11] J.A. DiMasi, "The Price of Innovation: New Estimates of Drug Development Costs," *Journal of Health Economics* 22 (2003) 151–185.

[12] J.A. DiMasi, "The Price of Innovation: New Estimates of Drug Development Costs," *Journal of Health Economics* 22 (2003) 151–185.

[13] Grid Technology Partners customer interview, March 2003.

[14] A case in point to illustrate this is e-mail. It is estimated that an employee may spend up 4 hours a day just managing and processing e-mail.

Examples from the Field - I

Area: Electronic Design Automation
Benchmark Application(s): 200 regression suites of Synopsys VCS jobs

Pre-Grid Deployment:
System: Sun 450 SMP, 2 x 480 Mhz Processors
System Cost: US $31,000
Processing Time: 288 hours

Post Grid Deployment:
System: Existing 50 single CPU, 1 Ghz desktop grid
System Cost: US $0
Grid Software: US $20,000
Processing Time: 12 hours

Metrics: 95% Faster, 35% cheaper

FIGURE 3.1   Productivity gains through Grid Computing–I.

relatively easy to determine the reduction in processing time due to increased computational capacity offered by the grid and the resulting emancipation of their employees' time. Along with the Grid Computing value elements discussed earlier in the section, productivity gains remain a strong driver for grid deployments.

Examples from the Field - II

Cognigen Grid Deployment: 2 X increase in computation power at 90% of cost
GlobeExplorer Grid Deployment: Reduced image processing time from 1 terabyte a month to 1 terabyte a week
Novartis Grid Deployment: Performed 3.18 years of Hidden Markov Models simulations in 7 days

---

FIGURE 3.2   Productivity gains through Grid Computing–II.

3.3 RISK ANALYSIS

In section 3.2, we discussed some of the areas to which Grid Computing adds immediate value for companies. In this section, we will evaluate the key risk factors that usually plague technology deployments and analyze the vulnerabilities of Grid Computing deployments.

3.3.1 Lock-in

Like most software (and hardware) vendors, Grid Computing vendors would probably prefer it if their software locked-in a customer for a recurring or a future revenue stream. However in the case of Grid Computing deployment, the IT manager will not be making any significant investments in *durable complementary assets*, which promote lock-in. Although cluster and SMP solutions will require significant investment in hardware, software, and supporting infrastructure and may contribute to customer lock-in. Customers should pay keen attention to which vendors are supporting the Grid Computing standards activities at the Global Grid Forum.

3.3.2 Switching Costs

Once a grid has been deployed, the primary switching cost will be driven by the effort required to integrate and enable enterprise applications to work on whatever replacement grid infrastructure that has been selected. This is usually performed by utilizing software development toolkits and is generally not too significant, but nonetheless should be carefully reviewed. Another way to mitigate switching costs is to introduce new grid software in the enterprise to support new grid-enabled applications, while letting the existing software deployment and its integration with legacy grid software remain unchanged.

3.3.3 Project Implementation Failure

The final risk factor is of project failure, either due to bad project management or incorrect needs assessment. One way to mitigate the risk of project failure is to take advantage of hosted pilot and professional services offered by grid software vendors. This will allow the IT manager to accurately pre-assess the suitability of the grid software, level of integration required, and feasibility (application speedup times, productivity gains, etc.). Hosted pilots are conducted solely on the vendors' data centers and have no impact on the operations of the company.

3.4 GRID MARKETPLACE

Grid Computing is an amalgamation of numerous technologies, used in harmony to provide a valuable and meaningful service to users. In this section, we will attempt to catalog different technologies into groups based on their role and function. Some companies may fall into more than one category. We identify key trends that we see developing in the marketplace.

3.4.1 Grid Taxonomy

We have defined a four-tier positioning matrix for *products* required to create a Grid Computing infrastructure. In addition, we have defined two vertical tiers for Grid Computing *services*. Some companies may have products that span all tiers of our matrix, while others may be specifically focused on a particular niche. It is important to realize that grouping products and services into categories in this manner is far from an exact science. As the Grid Computing market evolves, we expect to refine our assessments and add further categories (hence, more companies) to our matrix.

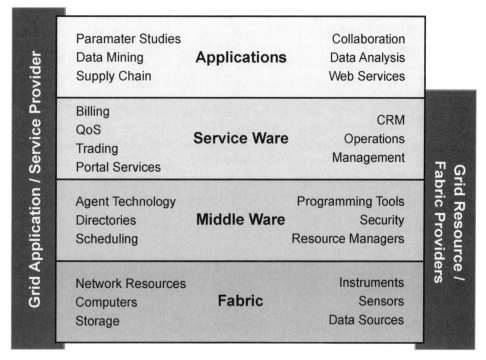

FIGURE 3.3   Grid Computing market taxonomy.

3.4.2 Fabric

The *fabric layer* in this taxonomy is similar to the fabric layer in the grid architecture model discussed in section 4.8, except that here we strictly refer to hardware products that are used to build the grid infrastructure. These include personal computers, clusters, high performance computers, and storage devices, as well as networking devices such as routers, switches, load balancers, caching systems, etc. The fabric layer also consists of various types of sensors (such as those discussed in Chapter 2) and other scientific instrumentation. Companies such as IBM, Hitachi, Fujitsu, Sun Microsystems, HP, Cisco Systems, and many others can be categorized as fabric providers in this taxonomy.

3.4.3 Middleware

The *middleware layer* consists of software vendors and products that manage and facilitate access to the resources available at the fabric layer. These products perform the function of resource management, security, scheduling, and execution of tasks. The middleware layer also includes the tools developed to facilitate use of resources by various applications. A *parallization* tool or compiler would reside in this layer of the matrix. Middleware companies include Avaki, United Devices, Entropia, GridSystems, Tsunami Research, and Platform Computing. Additionally, software companies that help in the development of applications for the grid, such as GridIron Software and Engineered Intelligence, also fall into this category.

3.4.4 Serviceware

The *serviceware layer* consists of vendors that provide the operational support systems for Grid Computing, or help integrate with existing systems. These include billing, account and management, and management software. Companies such as Gridfrastructure, GridXpert, and Montague River fall within the serviceware layer.

3.4.5 Applications

The *application layer* consists of all the software applications and vendors that will utilize the Grid Computing infrastructure. There are thousands of companies that will eventually adapt their applications for the grid. Some early adopters include Wolfram Research (makers of Mathematica), Lion Biosciences, Fluent Inc, Accerlys, Cadence, Mentor Graphics, and others.

3.4.6 Grid Resources Providers

The *Grid Resource Provider (GReP)* vertical consists of companies that provide the various resources listed in the fabric layer as services that can be leased on variable terms. For example, a customer may rent CPU cycles from a grid resource provider while running an especially compute intensive application, while at other times she may want to rent additional storage capacity. The GReP is not aware of the applications being executed by the customer on its platforms. The GReP buys equipment and software from fabric, middleware, and serviceware vendors. Grid Service Providers today include companies such as Gateway Corporation, ComputingX, IBM, and NTT Data.

3.4.7 Grid Applications Service Providers

The *Grid Applications Service Provider (GASP)* provides end-to-end Grid Computing services to the user of a particular application or applications. The customer in this case will purchase "application time" from the provider, and will provide the data or parameters to the GASP through an application portal and in the future through published Web service specifications. The GASP may choose to purchase services from the GReP or may choose to build the infrastructure organically.

3.4.8 Grid Consultants

Finally, there are consulting companies specializing in providing services to Grid Computing vendors as well as companies that want to grid-enable their applications and infrastructure. IBM, HP, and Ascential are some of the early software consulting firms emerging in this field.

3.5 CONCLUSION

Grid Computing is having a remarkable impact at enterprises in all business sectors. Companies are realizing that Grid Computing offers tremendous return on investment with fast payback and a drastically lower total cost of ownership as compared with other technology solutions. While building a business case for introducing Grid Computing into an enterprise, tangible as well as intangible gains should be considered. The vendor universe offering Grid Computing solutions and services is growing every day. All major information technology firms have wholeheartedly embraced the technology.

4 Grid Computing Technology–An Overview

Ahmar Abbas

Grid Technology Partners

In This Chapter

- History of Grid Computing
- Definition of Grid Computing
- A Grid Computing Model
- Grid Computing Protocols
- Types of Grids

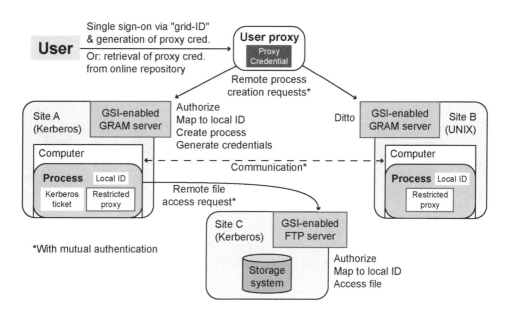

4.1 INTRODUCTION

Grid Computing, as we know it today, is the evolution and amalgamation of numerous development efforts that have been going on for many years. This chapter provides a brief historical background of Grid Computing and some of the other similar technologies that preceded it. This is followed by detailed discussions on the model, protocols, and architecture of Grid Computing.

4.2 HISTORY

In the early-to-mid 1990s, there were numerous research projects underway in the academic and research community that were focused on distributed computing. One key area of research focused on developing tools that would allow distributed high performance computing systems to act like one large computer. At the IEEE/ACM 1995 Super Computing conference in San Diego, 11 high speed networks·were used to connect 17 sites with high-end computing resources for a demonstration to create one super "metacomputer." This demonstration was called I-Way and was led by Ian Foster of the United States Department of Energy's Argonne National Labs and University of Chicago.[1] Sixty different applications, spanning various faculties of science and engineering, were developed and run over this demonstration network. Many of the early Grid Computing concepts were explored in this demonstration as the team created various software programs to make all computing resources work together.

The success of the I-Way demonstration led the United States government's DARPA agency, in October 1996, to fund a project to create foundation tools for distributed computing. The research project was led by Ian Foster of ANL and Carl Kesselman of University of Southern California. The project was named *Globus* and the team created a suite of tools that laid the foundation for Grid Computing activities in the academic and research communities. At the 1997 Super Computing Conference, 80 sites worldwide running software based on the Globus Toolkit were connected together.

This effort started to be referred to as Grid Computing—coined to play on the analogy to the electrical power grid.[2] Grid computing would make tremendous computing power available to anybody, at anytime, and in a truly transparent

---

[1] Foster, Ian, 2000, "Internet Computing and the Emerging Grid," *Nature,*12/2002.

[2] Waldrop, Mitchel M., 2002, "Grid Computing," *Technology Review* 3/2002.

manner, just as, today, the electric power grid makes power available to billions of electrical outlets.

Grid Computing in the academic and research communities remained focused on creating an efficient framework to leverage distributed high-performance computing systems. But with the explosion of the Internet and the increasing power of the desktop computer during the same period, many efforts were launched to create powerful distributed computing systems by connecting together PCs on the network. In 1997, Entropia was launched to harness the idle computers worldwide to solve problems of scientific interest. The Entropia network grew to 30,000 computers with aggregate speed of over one teraflop per second. A whole new field of philanthropic computing (discussed in section 4.17) came about in which ordinary users volunteered their PCs to analyze research topics such as patient's response to chemotherapy, discovering drugs for AIDS, and potential cures for anthrax.

Although none of the above projects could be successfully monetized by companies, they did, however, attract a lot more media attention than any of the earlier projects in the academic and research world. Starting in late 2000, articles on Grid Computing moved from the trade press to the popular press. In rapid fire succession articles appeared in, for example, *New York Times, Economist, Business 2.0, Red Herring, Washington Post, Financial Times, Yomiuri Shimbum, The Herald* of Glasgow, *Jakarta Post, Dawn* of Karachi, etc. In fact, the trend has only accelerated recently. An analysis of Lexus-Nexis data shows that Grid Computing references in popular U.S. media is up dramatically to 500 citations from 100 in Q4 of 2001.[3] Grid Technology Partners' Grid Looking Glass Index (see Figure 4.1) which tracks Grid Computing related searches on the Google search engine tripled in the first year of its inception.

Today, large corporations such as IBM, Sun Microsystems, Intel, Hewlett Packard, and some smaller companies such as Platform Computing, Avaki, Entropia, DataSynapse, and United Devices are putting their marketing dollars to good use and creating the next generation of thought-leadership around Grid Computing that is focused on business applications rather than academic and basic research applications.

---

[3] Grid Technology Partners, Lexis-Nexis search, April 2002.

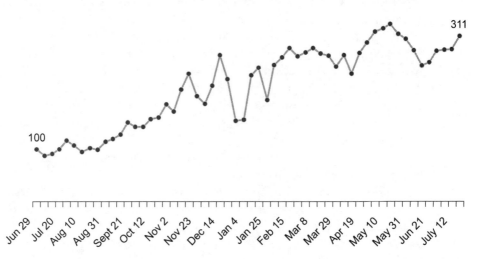

FIGURE 4.1   Grid Looking Glass Index.

4.3 HIGH-PERFORMANCE COMPUTING

High-performance computing generally refers to what has traditionally been called supercomputing. There are hundreds of supercomputers deployed throughout the world. Key parallel processing algorithms have *already* been developed to support execution of programs on different, but co-located processors. High-performance computing system deployment, contrary to popular belief, is not limited to academic or research institutions. In fact, more than half of supercomputers deployed in the world today are in use at various corporations.[4]

The industries in which high performance systems are deployed are numerous in nature. Table 4.1 shows the distribution of the top 500 supercomputers by their industries.

As mentioned in the preceding section, it was the desire to share high-performance computing resources amongst researchers that led to the development of Grid Computing technology and some of its fundamental infrastructure. This is discussed extensively in Chapter 7.

High-performance computing statistics are extremely important because they tell us which industries already have demand for tremendous computing power.

---

[4] Information on the world's top 500 supercomputers is compiled twice a year jointly by the University of Mannheim and the University of Tennessee. It can be found at *www.top500.org*

TABLE 4.1 HPC Deployment by Industry

Industry Area	Sample Companies
Telecommunication	Sprint, Duetsche Telekom
Finance	Charles Schwab
Automotive	BMW, GM
Database	State Farm, Starbucks
Transportation	Oy Saimaa Lines
Electronics	Cisco, Motorola
Geophysics	Aramco, Shell
Aerospace	Dassault Avaition
Energy	Centrica
Worldwide Web	Newsky, Amazon
Information Services	EDS
Chemistry	Bayer
Manufacturing	Alcoa
Mechanics	Hitachi
Pharmaceutics	Aventis Pharma

Source: University of Mannhiem

4.4 CLUSTER COMPUTING

Cluster computing came about as a response to the high prices of supercomputers, which made those systems out of reach for many research projects. Clusters are high-performance, massively parallel computers built primarily out of commodity hardware components, running a free-software operating system such as Linux or FreeBSD, and interconnected by a private high-speed network. It consists of a cluster of PCs, or work-stations, dedicated to running high-performance computing tasks. The nodes in the cluster do not sit on users' desks, but are dedicated to running cluster jobs. A cluster is usually connected to the outside world through only a single node.

Cluster computing has been around since 1994 when the first Beowulf clusters were developed and deployed.[5] Since then, numerous tools have been developed to

---

[5] Beowulf clusters were developed by Thomas Sterling and Don Becker while working at the Center of Excellence in Space and Data and Information Sciences, a division of University Space Research Association located at NASA Goddard Space Flight Center.

run and manage clusters. Platform Computing, a firm that is now a leader in Grid Computing, developed many of the early load-balancing tools for clusters. Additionally, tools have also been developed to adapt applications run in the parallel cluster environment. One such tool, ForgeExplorer, can check if particular applications are suitable for parallelization and determine if they would be suitable to run on a cluster.

The exponential growth in microprocessor speeds over the last decade has now made it possible to create truly impressive clusters. This year an AMD® Athlon™-based cluster at University of Heidelberg in Germany was tested at 825 Gflops, making it the 35th fastest high performance computer in the world.[6] Clusters are widely deployed in industries such as life sciences, digital entertainment, finance, etc.

IEEE Computer Society Task Force on Cluster Computing conducts a yearly conference that is designed to bring together international cluster and Grid Computing researchers, developers, and users to present and exchange the latest innovations and findings that drive future research and products. It is expected that the Grid Computing community will benefit from the years of experience that the cluster community has in building tools that allow applications to share distributed computing resources. Cluster computing and cluster-based grids are discussed in chapter 6.

4.5 PEER-TO-PEER COMPUTING

Although the recent growth of user-friendly file-sharing networks, such as Napster or Kazaa, has only now brought Peer-to-Peer (P2P) networks and file sharing into the public eye, methods for transferring files and information between computers have been, in fact, around almost as long as computing itself. Until recently, however, systems for sharing files and information between computers were exceedingly limited. They were largely confined to Local Area Networks (LANs) and the exchange of files with known individuals over the Internet. LAN transfers were executed mostly via a built-in system or network software while Internet file exchanges were mostly executed over an FTP (File Transfer Protocol) connection. The reach of this Peer-to-Peer sharing was limited to the circle of computer users an individual knew and agreed to share files with. Users who wanted to communicate with new or unknown users could transfer files using IRC (Internet Relay Chat) or other similar bulletin boards dedicated to specific subjects, but these methods never gained mainstream popularity because they were somewhat difficult to use.

Today, there are a number of advanced P2P file sharing applications, and the reach and scope of peer networks have increased dramatically. The two main mod-

---

[6] The AMD-based cluster was tested based on Linpack test performed by Top500.org.

els that have evolved are the centralized model, such as the one used by Napster, and the decentralized model like the one used by Gnutella.

In the centralized model of P2P, file sharing is based around the use of a central server system that directs traffic between individual registered users. The central servers maintain directories of the shared files stored on the respective PCs of registered users of the network. These directories are updated every time a user logs on or off the Napster server network. Each time a user of a centralized P2P file sharing system submits a request or searches for a particular file, the central server creates a list of files matching the search request by cross-checking the request with the server's database of files belonging to users who are currently connected to the network. The central server then displays that list to the requesting user. The requesting user can then select the desired file from the list and open a direct HTTP link with the individual computer that currently possesses that file. The download of the actual file takes place directly, from one network user to the other. The actual file is never stored on the central server or on any intermediate point on the network.

The decentralized model of P2P file sharing does not use a central server to keep track of files. Instead, it relies on each individual computer to announce its existence to a peer, which in turn announces it to all the users that it is connected to, and so on. The search for a file follows a similar path. If one of the computers in the peer network has a file that matches the request, it transmits the file information (name, size, etc.) back through all the computers in the pathway to the user that requested the file. A direct connection between the requester and the owner of the file is directly established and the file is transferred.

4.6 INTERNET COMPUTING

The explosion of the Internet and the increasing power of the home computer prompted computer scientists and engineers to apply techniques learned in high-performance and cluster-based distributed computing to utilize the vast processing cycles available at users' desktops. This has come to be known as Internet computing.

Large compute intensive projects are coded so that tasks can be broken down into smaller subtasks and distributed over the Internet for processing. Volunteer[7] users then download a lightweight *client* onto their desktop, which periodically communicates with the central server to receive tasks. The client initiates the tasks only when the desktop CPU is not in use. Upon completion of the task, it communicates results back to the central server. The central server aggregates the information received from all the different desktops and compiles the results.

---

[7] In section 4.17.3 we classify some of the Internet computing projects as an emerging area called philanthropic computing.

United Devices, Entropia, and others have established large groups of users that volunteer their desktops for large computing projects.

Table 4.2 lists Internet computing projects broken down by area of interest. This is to emphasize the point that while some Internet computing projects have gotten a lot of press recently, there are many more that are working quietly in the background.

TABLE 4.2 Internet Computing Projects

Area of Interest	Projects
Science	SETI@Home Analytical Spectroscopy Research Group Evolutionary Research eOn Climateprediction.com Distributed Particle Accelerator Design
Life Sciences	Folderol Folding@Home Genome@Home FightAIDS@Home Übero Drug Design Optimization Lab The Virtual Laboratory Project Distributed Folding Community TSC Find-a-Drug
Cryptography	Distributed.net ECCp-109
Mathematics	Great Internet Mersenne Prime Search Proth Prime Search ECMNET n!+1 and n!-1 Prime Search Minimal Equal Sums of Like Powers GRISK MM61 Project 3x + 1 Problem Pi(x) Project Distributed Search for Fermat Number Divisors PCP@Home Generalized Woodall Numbers Generalized Fermat Prime Search ZetaGrid Strong Pseudoprime Search Wilson Prime Search Largest Proth Prime Search Search for Primes of the Form k.2n - 1 Weiferich Prime Number Search Seventeen or Bust Factorizations of Cyclotomic Numbers

Source: Grid Technology Partners

Internet computing projects, although not profitable, have allowed companies to understand large-scale distributed computation projects. Many of these companies, which were originally funded to harness the power of the consumers' desktops connected to the Internet, are retooling their products for enterprise applications.

4.7 GRID COMPUTING

Grid computing tries to bring, under one definitional umbrella all the work being done in the high performance, cluster, peer-to-peer, and Internet computing arenas. Coming up with a definition for Grid Computing, therefore is not as easy as one would have expected. Vendors, academics, trade, as well as the popular press have all tried to define Grid Computing.

Some of the definitions of Grid Computing that we have uncovered include:

■ The flexible, secure, coordinated resource sharing among dynamic collections of individuals, institutions, and resources.
■ Transparent, secure, and coordinated resource sharing and collaboration across sites.
■ The ability to form virtual, collaborative organizations that share applications and data in an open heterogeneous server environment in order to work on common problems.
■ The ability to aggregate large amounts of computing resources which are geographically dispersed to tackle large problems and workloads as if all the servers and resources are located in a single site.
■ A hardware and software infrastructure that provides dependable, consistent, pervasive, and inexpensive access to computational resources.
■ The Web provides us information—the grid allows us to process it.

Although some of these definitions are quite good, the following broader definition of Grid Computing serves the purpose of this book more fully and will be used to describe grid systems:

Grid Computing enables virtual organizations *to share geographically distributed* resources *as they pursue common goals, assuming the* absence *of central location, central control, omniscience, and an existing trust relationship.*[8]

---

[8] Training presentation. *www.globus.org*

Virtual organizations can span from small corporate departments that are in the same physical location to large groups of people from different organizations that are spread out across the globe. Virtual organizations can be large or small, static or dynamic. Some may come together for a particular event and then be disbanded once the event expires.

Some examples of a virtual organization are:

1. Boeing's Blended Wing Body design team located in numerous Boeing offices around the world.
2. Worldcom's Global VPN Product Management team with members in 28 countries working on defining product specifications.
3. An accounting department of a company.
4. An emergency response team created to tackle an oil spill in the Gulf of Mexico.

A *resource* is an entity that is to be shared. It can be computational such as a personal digital assistant, laptop, desktop, workstation, server, cluster, and supercomputer or a storage resource such as a hard drive in a desktop, (Redundant Array of Inexpensive Disks), and terabyte storage device. Sensors (of the type discussed in Chapter 1) are another type of resource. Bandwidth is yet another resource that is used in the activities of the virtual organization.

Absence of a central location and central control implies that grid resources do not require a particular central location for their management. The final key point is that in a grid environment the resources do not have prior information about each other nor do they have pre-defined security relationships.

We mentioned earlier that this is a broad and all-encompassing definition of Grid Computing. There will be degrees to which certain grid deployments and grid products meet or do not meet the above criteria.

4.7.1 Peer-to-Peer Networks and Grid Computing

Peer-to-peer networks (e.g., Kazaa) fall within our definition of Grid Computing.[9] The resource in peer-to-peer networks is the storage capacity of each (mostly desktops) node. Desktops are globally distributed and there is no central controlling authority. The exchange of files between users also does not predicate any pre-existing trust relationship. It is not surprising, given how snugly P2P fits in our definition

---

[9] Kazaa is the latest free file sharing product. It has had almost 100,000,000 downloads to date, with an average of two million new downloads a week.

of Grid Computing, that the Peer to Peer Working Group has become part of the grid standards body, the Global Grid Forum (GGF). The GGF and its role will be discussed in section 4.12.

4.7.2 Cluster Computing and Grid Computing

From a Grid Computing perspective, a cluster is a resource that is to be shared. A grid can be considered a cluster of clusters.[10]

4.7.3 Internet Computing and Grid Computing

Internet computing examples presented earlier in our opinion fit this broad definition of Grid Computing. A virtual organization is assembled for a particular project and disbanded once the project is complete. The shared resource, in this case, is the Internet connected desktop.

4.8 GRID COMPUTING MODEL

In the previous section, we presented a definition of Grid Computing. But before laying out the architectural model that supports the definition, it is important to understand the key challenges that need to be addressed. One way to look at what a grid or grid system is trying to accomplish is to imagine a grid implementation that is trying to run program X using resources at site Y, subject to virtual community policy P, providing access to data at Z according to policy Q.

Any effort to accomplish the above runs into two classes of problems. First, and foremost, the application, or program X, has to be able to work in an environment Y that could be heterogeneous and geographically dispersed (work in a parallel computing paradigm). We will push out the discussion of programming applications that can take advantage of grids to section 4.15 and 4.16 of this chapter and focus on the second problem, or the systems problem. The systems problem is figuring out how to coordinate the use of the resources at sites Y and Z under the various restrictions on their usage as defined by policies P and Q. In other words, who gets to use what, when, and why?

Although coordinated use of resources is not a trivial problem in a closed environment, it gets more complicated when it is attempted across geographical and organizational boundaries. Some of the key questions that come up when sharing resources across boundaries are:

---

[10] We were introduced to this definition by Ian Baird, chief strategist at Platform Computing.

- Identity and Authentication—Is this user who he says he is? Is this program the right program?
- Authorization and Policy—What can the user do on the grid? What can the application do on the grid? What resources are the user and or application allowed to access?
- Resource Discovery—Where are the resources?
- Resource Characterization—What types of resources are available?
- Resource Allocation—What policy is applied when assigning the resources? What is the actual process of assigning the resources. Who gets how much?
- Resource Management—Which resource can be used at what time and for what purpose?
- Accounting/Billing/Service Level Agreement (SLA)—How much of the resources is being used? What is the rating schedule? What is the SLA?
- Security—How do I make sure that this is done securely? How do we know if we have been compromised? What steps are taken once a security breach is detected?

To overcome the systems problem, a set of protocols and mechanisms need to be defined that address the security and policy concerns of the resource owners and users. The grid protocol(s) should be flexible enough to deal with many resource types, scale to large numbers of resources with many users and many program components. More important, it should do all the above in an efficient and cost effective manner. In addition to the grid protocols that have to be defined, a set of grid applications programming interfaces (APIs) and software development toolkits (SDKs) need to be defined. They provide interfaces to the grid protocols and services as well as facilitate application development by supplying higher-level abstraction.

The model, grid protocols, and accompanying APIs and SDKs have been hugely successful in the Internet world. The grid architecture model shown in Figure 4.2 has been closely aligned with the Internet protocol architecture as defined by the Open Systems Interconnect (OSI) Internet stack.

Protocols, services, and APIs occur at each level of the grid architecture model. Figure 4.3 shows the relationship between APIs, services, and protocols. At each protocol layer in the grid architecture, one or more services are defined. Access to these services is provided by one or more APIs. More sophisticated interfaces, or software development toolkits, provide complex functionality that may not map one to one onto service functions and may combine services and protocols at lower levels in the grid protocol stack.

At the top of this figure, we include languages and frameworks, which utilize the various APIs and SDKs to provide programming environments to the grid application.

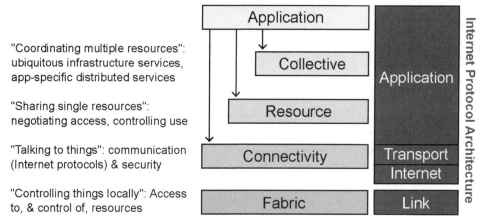

FIGURE 4.2 Grid computing architecture model.

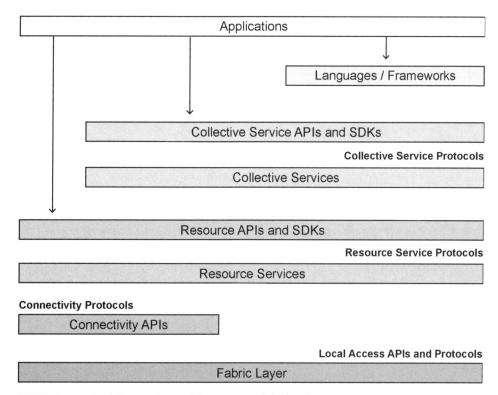

FIGURE 4.3 Grid Computing architecture model—detail.

The *fabric layer* includes the protocols and interfaces that provide access to the resources that are being shared. We have already identified these earlier as compute resources, data resources, etc. This layer is a logical view rather than a physical view. For example, the view of a cluster with a local resource manager is defined by the local resource manager and not the cluster hardware. Likewise, the fabric provided by a storage system is defined by the file system that is available on that system and not the raw disk or tapes.

The *connectivity layer* defines core protocols required for grid-specific network transactions. These utilize the existing Internet protocols such as IP, Domain Name Service, various routing protocols such as BGP, and so on. Another set of protocols defined by the connectivity layer include the core grid security protocol. This is also known as the Grid Security Infrastructure (GSI). GSI provides uniform authentication, authorization, and message protection mechanisms. It also provides for a single sign-on to all the services that will be used and it utilizes public key technology such as X.509.[11]

The *resource layer* defines protocols required to initiate and control sharing of *local* resources. Protocols defined at this layer include:

- Grid Resource Allocation Management (GRAM)—Remote allocation, reservation, monitoring, and control of resources
- GridFTP (FTP Extensions)—High performance data access and transport
- Grid Resource Information Service (GRIS)—Access to structure and state information

These protocols are built on the connectivity layer's grid security infrastructure and utilize standard IP protocols for communications.

The *collective layer* defines protocols that provide system oriented (versus local) capabilities for wide scale deployment. This includes index or meta-directory services so that a custom view can be created of the resources available on the grid. It also includes resource brokers that discover and then allocate resources based on defined criteria.

The *application layer* defines protocols and services that are targeted toward a specific application or a class of applications. This layer is currently the least defined in the grid architecture.

In short, each layer provides a set of services that allow Grid Computing resources to be identified and accessed securely based on a set of rules. The rules

---

[11] Supporting infrastructure such as certificate authorities and certificate and key management systems is also required.

are defined both by the user of the resource and the owner. The services can be accessed by programmers through a set of applications programming interfaces and software development toolkits that have been defined for each layer.

4.9 GRID PROTOCOLS

In the preceding section, protocols associated with each layer in the grid architecture were discussed. In this section we address each of these protocols individually:

- Grid Security Infrastructure
- Grid Resource Allocation Management
- Grid File Transfer Protocol
- Grid Information Services

4.9.1 Security: Grid Security Infrastructure

It is safe to say that the way security is handled in grids will ultimately be the single most important determinant of its mainstream adoption and deployment. It is, therefore, not surprising that a significant amount of effort is being focused on grid security by the standards body and vendors in this space.

Security is defined in the resource layer of the grid architecture. It is important because the resources being used may be valuable and the problems being solved or tasks being attempted sensitive. The security problem in a grid environment is complex because resources are often located in different administrative domains with each resource potential having its own policies and procedures.

Security concerns are further complicated by the fact that there are different requirements by users, resource owners, and developers who are creating or adapting their current products and tools to take advantage of the grid technology.

The user's (person or another program) expectations are that a secure grid system will be easy to use, provide single sign-on capability, allow for delegation, and support all key applications.

The resource owners require that security should specify local access control, have robust and detailed auditing and accounting, and should be able to integrate with local security infrastructure. There should be protection in the event other resources get compromised.

From a developer's standpoint, the grid security protocol should have a robust API/SDK that allows direct calls to the various security functions.

The Grid Security Infrastructure (GSI) for grids has been defined by creating extensions to standard and well-known protocols and APIs. Extensions for Secure Socket Layer/ Transport Layer Security (SSL/TLS) and X.509 have been defined to allow single sign-on (proxy certificate) and delegation. We will not go into the details of X.509 in this chapter.[12]

The X.509 proxy certificate grid extension defines how a short-term, restricted credential can be created from a normal, long-term X.509 credential. This supports single sign-on and delegation through "impersonation" and is also an Internet Engineering Task Force (IETF) draft.

The Generic Security Service (GSS) API extensions have been created and are under review at the Global Grid Forum. GSS is an IETF standard that provides functions for authentication, delegation, and message protection. Figure 4.4 shows the Grid Security Infrastructure in action. The request submitted is as follows: "Create processes at A and B that Communicate & Access Files at C."

GSI has been implemented at numerous sites. In fact, almost all of the research and academic grid activities use GSI. The Globus Certificate authority alone has is-

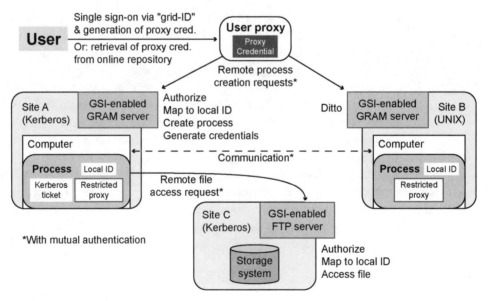

FIGURE 4.4   GSI in action.

---

[12] For more information on X.509 please refer to the following Internet Engineering Task Force (IETF) site: *http://www.ietf.org/html.charters/pkix-charter.html*

sued over 4000 user and host certificates. The standardization process for the GSI has begun at the Global Grid Forum.

4.9.2 Resource Management: Grid Resource Allocation Management Protocol

The Grid Resource Allocation and Management protocol and client API allows programs to be started on remote resources. A Resource Specification Language (RSL) has been developed as a common notation for exchange of information between applications, resource brokers, and local resource managers. RSL provides two types of information:

- Resource requirements: machine type, number of nodes, memory, etc.
- Job configuration: directory, executable, arguments, environment

    An example of an RSL-based requirement would be as follows:

    "create 5-10 instances of *myprog*, each on a machine with at least 64MB memory that is available to me for 4 hours, or 10 instances, on a machine with at least 32MB of memory"

    GRAM protocol is a simple, HTTP-based remote procedure call (RPC). It sends messages such as job request, job cancel, status, and signal (see Figure 4.5). Event notifications for state changes include pending, active, done, failed, or suspended.

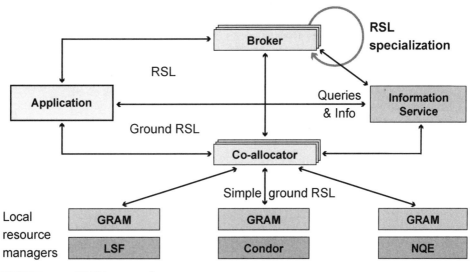

FIGURE 4.5 GRAM protocol.

GRAM-2 protocol includes multiple resource types, such as storage, network, sensors, etc. It will also use Web Services protocols such as Web Services Definition Language (WSDL) and Simple Object Access Protocol (SOAP).

4.9.3 Data Transfer: Grid File Transfer Protocol

There are numerous examples of grids today that have to perform sophisticated, computationally intensive analyses on petabytes of data. In these examples, data are being collected at one location while the researchers who need access to the data are distributed across the globe. One of the key requirements for these data-intensive grids is high-speed and reliable access to remote data. The standard FTP protocol has been extended while preserving interoperability with existing servers to develop GridFTP. The extensions provide for striped/parallel data channels, partial files, automatic and manual TCP buffer size settings, progress monitoring, and extended restart functionality.

The protocol extension to FTP for the grid (GridFTP) has been submitted as a draft to the Global Grid Forum Data Working group.

4.9.4 Information Services: Grid Information Services

A set of protocols and APIs are defined in the resource layer that provides key information about the grid infrastructure. Grid Information Service (GIS) provides access to static and dynamic information regarding a grid's various components and includes the type and state of available resources.

There are two types of Grid Information Services. The Grid Resource Information Service (GRIS) and the Grid Index Information Service (GIIS). The GRIS supplies information about a specific resource while the GIIS is an aggregate directory service. GIIS provides a collection of information that has been gathered from multiple GRIS servers.

The Grid Resource Registration protocol is used by resources to register with the GRIS servers. The Grid Resource Inquiry protocol is used to query a resource description server for information and also query the aggregate server for information.

4.10 GLOBUS TOOLKIT

Globus™ is a *reference implementation* of the grid architecture and grid protocols discussed in the preceding sections. Globus is a United States government-funded project that provides software tools that make it easier to build grids and grid-based applications. These tools are collectively called the Globus Toolkit™. The Globus Toolkit is an open architecture, open source software toolkit. Many projects and developers around the world have contributed to the Globus Toolkit. A growing number of companies have committed to supporting this open source activity by, for example, porting the software to their platforms. The main research teams are located at Argonne National Labs, University of Chicago, NCSA, and University of Southern California.

The Globus Toolkit includes tools and libraries for solving problems in the following areas:

- Security—Supports GSI
- Resouce Management—Supports GRAM. It is implemented as a component called Gatekeeper.
- Data Management—Supports GridFTP as well as replica services.
- Information Services—Supports GIS.

The recently released Globus Toolkit version 3.0 supports the following platforms:

- Linux Kernel 2.x™, Intel x86™
- Linux Kernel 2.4™, Intel IA-64™ (Itanium™)
- IRIX 6.5™, MIPS™
- Solaris 2.8™, UltraSPARC™
- AIX 5.1™
- Compaq Tru64™

The Globus Toolkit is available free of charge from the Globus site at *www.globus.org*. A commercially supported version, Platform Globus, is available from Platform Computing.

4.11 OPEN GRID SERVICES ARCHITECTURE

Open Grid Services Architecture, an effort led by IBM and the Globus team, tries to marry the Web Services architecture with the Grid Computing architecture. Taking advantage of the experience gained from the Globus implementation of grid technologies and Web Services standards, OGSA will model traditional grid resources (computation, storage) as a Grid Service. OGSA was first presented at the Global Grid Forum IV in Toronto, Canada in February, 2002. The initial objectives were first outlined in "The Physiology of the Grid—An Open Grid Services Architecture for Distributed Systems Integration."[13]

The effort is based on the underlying similarities between what grid technologies and Web Services have been trying to accomplish, albeit on separate tracks: the sharing of resources and facilitating the creation of virtual organizations. In the case of Web Services, this includes sharing of business logic, data, and processes amongst external e-business partners (a type of virtual organization). In the case of grids, the virtual organization is sharing computation and database resources among a team that has been specifically created to tackle a particular scientific or engineering problem. Both virtual organizations are unlimited by physical location. One major difference is that Web Services address persistent services while grids must also support transient services. An example of a transient service would be the invocation of a video conference resource and its subsequent teardown once the activity is completed.

The recently released Grid Service Specification provides detailed specification for the conventions that govern:[14]

- How Grid Services are created and discovered
- How Grid Service instances are named and referenced
- Interfaces that define any Grid Service

There is still a lot of work that needs to be done in expanding the above specification. Whether OGSA is an IBM-driven marketing push to counter Microsoft's .NET™ initiative, or whether it is a serious contender that will be heartily accepted by enterprises, remains to be seen. There is, however, great optimism that OGSA will facilitate adoption of grid technologies for traditional IT applications in addition to the R&D applications because it is based on standard Web Services standards. Almost all the major grid technologies vendors have signed on to support OGSA and there has been no competing effort put forth at the Global Grid Forum.

4.12 GLOBAL GRID FORUM

The Global Grid Forum is the main standards body governing the grid community. The functioning of the organization is modeled around other standards bodies, notably the Internet Engineering Task Force.

---

[13] Ian Foster, Carl Kesselman, Jeffrey Nick, and Steven Tuecke.

[14] Tuecke, Czajkowski, Foster, Frey, Graham, and Kessleman, "Grid Service Specification" 6/13/2002.

The Global Grid Forum is the result of a merger between the Grid Forum, eGrid European Grid Forum, and the Asia-Pacific grid community. In April 2002, the New Productivity Initiative—which was formed in 2000 to create a layered, open-API specification for Distributed Resource Management (DRM) by documenting specifications and standards that allow and promote interoperability—merged with the Global Grid Forum. Also in April 2002, the Peer-to-Peer Working Group—which formerly created best practices which enabled interoperability between computing and networking systems for the peer-to-peer community—merged with the Global Grid Forum.

The mission of the Global Grid Forum is:

> to focus on the promotion and development of Grid technologies and applications via the development and documentation of "best practices," implementation guidelines, and standards with an emphasis on "rough consensus and running code."[15]

The work of the Global Grid Forum is performed within its various working groups and research groups. A working group is generally focused on a very specific technology or issue with the intention to develop one or more specific documents aimed generally at providing specifications, guidelines, or recommendations. A research group is often longer-term focused, intending to explore an area where it may be premature to develop specifications. Table 4.3 lists some of the current working groups at the Global Grid Forum.

The Global Grid Forum meets three times a year. There has been, not surprisingly, a steady increase in attendees at these meetings

TABLE 4.3 Global Grid Forum Working Groups[16]

Area	Working Group	Research Group
Grid Information Services / Performance (GIS-PERF)	Grid Object Specification (GOS) Grid Notification Framework (GNF) Metacomputing Directory Services (MDS) Grid Monitoring Architecture (GMA)	
Grid Information Services / Performance (GIS-PERF)	Grid Object Specification (GOS) Grid Notification Framework (GNF) Metacomputing Directory Services (MDS) Grid Monitoring Architecture (GMA) Network Measurement (NM) Discovery and Monitoring Event Description (DAMED)	Relational Database Information Services (RDIS)

(Continues)

---

[15] Global Grid Forum, *www.globalgridforum.org*

[16] Global Grid Forum, *www.globalgridforum.org*

TABLE 4.3 Global Grid Forum Working Groups (*continued*)

Area	Working Group	Research Group
Security (SEC)	Grid Security Infrastructure (GSI) Grid Certificate Policy (GCP)	
Scheduling (SCHED)	Distributed Resource Management Application API Working Group (DRMAA) Scheduling Dictionary (DICT) Scheduler Attributes (SA)	
Architecture (ARCH)	JINI NPI OGSI	Grid Protocol Architecture (GPA) Accounting Models (ACCT)
Data	GridFTP Data Access and Integration Services	Data Replication (REPL) Persistent Archives (PA)
Applications, Programming Models, Environments (APME)		Applications and Testbeds (APPS) Grid User Services (GUS) Grid Computing Environments (GCE) Advanced Programming Models (APM) Advanced Collaborative Environments (ACE)
Peer-to-Peer	NAT/Firewall Taxonomy Peer-to-Peer Security File Services Trusted Library	

Source: Grid Technology Partners

4.13 TYPES OF GRIDS

Grid computing vendors have adopted various nomenclatures to explain and define the different types of grids. Some define grids based on the structure of the *organization* (virtual or otherwise) that is served by the grid, while others define it by the principle *resources* used in the grid. In this section we attempt to classify these varying definitions.

4.13.1 Departmental Grids

Departmental grids are deployed to solve problems for a particular group of people within an enterprise. The resources are not shared by other groups within the enterprise. Following is a list of vendor definitions that we believe refer to departmental grids.

Cluster Grids: Cluster grid is a term used by Sun Microsystems and consists of one or more systems working together to provide a single point of access to users.[17] It is typically used by a team for a single project and can be used to support both high throughput and high performance jobs.

Infra Grids: Infra grid is a term used by IBM to define a grid that optimizes resources within an enterprise and does not involve any other internal partner. It can be within a campus or across campuses.[18]

4.13.2 Enterprise Grids

Enterprise grids consist of resources spread across an enterprise and provide service to all users within that enterprise. The following vendor definitions fall into this category.

Enterprise Grids: An enterprise grid, according to Platform Computing, is deployed within large corporations that have a global presence or a need to access resources outside a single corporate location. Enterprise grids run behind the corporate firewall.

Intra Grids: According to IBM, resource sharing among different groups within an enterprise constitutes an intra grid. An intra grid can be local or traverse the wide area network. Intra grids are located within the corporate firewall.

Campus Grids: Campus grids, according to Sun Microsystems, enable multiple projects or departments to share computing resources in a cooperative way. Campus grids may consist of dispersed workstations and servers as well as centralized resources located in multiple administrative domains, in departments, or across the enterprise.

4.13.3 Extraprise Grids

Extraprise grids are established between companies, their partners, and their customers. The grid resources are generally made available through a virtual private network. Following are some of the terms used by various vendors to describe such grids.

Extra Grids: Extra grids, according to IBM, enable sharing of resources with external partners. This assumes that connectivity between the two enterprises

---

[17] Please see *www.sun.com* for more details about their grid categories.

[18] Adams, John, IBM E-Technology Center, Austin, TX.

is through some trusted service, such as a private network or a virtual private network.

Partner Grids: Platform Computing defines these as grids between organizations within similar industries, which have a need to collaborate on projects and use each other's resources as a means to reach a common goal.

4.13.4 Global Grids

Grids established over the public Internet constitute *global grids*. They can be established by organizations to facilitate their business or purchased in part, or in whole, from service providers. Following are some vendor definitions that fall in this category.

Global Grids: Global grids, as defined by Sun, allow users to tap into external resources. Global grids provide the power of distributed resources to users anywhere in the world for computing and collaboration. They can be used by individuals or organizations to send overflow work over the public network to a grid services provider.

Inter Grids: Inter grids, according to IBM, provide the ability to share compute and data/storage resources across the public Web. This can involve sharing resources with other enterprises or buying or selling of excess capacity.

4.13.5 Compute Grids

Compute grids are created solely for the purpose of providing access to computational resources. Compute grids can be further classified by the type of computational hardware deployed.

Desktop Grids: These are grids that leverage the compute resources of desktop computers. Because of the true (but unfortunate) ubiquity of Microsoft® Windows® operating system in corporations, desktop grids are assumed to apply to the Windows environment. The Mac OS™ environment is supported by a limited number of vendors.

Server Grids: Some corporations, while adopting Grid Computing , keep it limited to server resources that are within the purview of the IT department. Special servers, in some cases, are bought solely for the purpose of creating an internal "utility grid" with resources made available to various departments. No desktops are included in server grids. These usually run some flavor of the Unix/Linux operating system.

High-Performance/Cluster Grids: These grids constitute high-end systems, such as supercomputers or HPC clusters. These are discussed in more detail in later chapters.

4.13.6 Data Grids

Grid deployments that require access to, and processing of, data are called *data grids*. They are optimized for data-oriented operations. Although they may consume a lot of storage capacity, these grids are not to be confused with storage service providers.

4.13.7 Utility Grids

We define *utility grids* as being commercial compute resources that are maintained and managed by a service provider. Customers that have the need to augment their existing, internal computational resources may purchase "cycles" from a utility grid. In addition to overflow applications, customers may choose to use utility grids for business continuity and disaster recovery purposes. In our grid taxonomy (Chapter 3.4), utility grid providers are also called Grid Resource Providers (GReP). Along with computing resources, some utility grids also offer key business applications that can be purchased "by the minute."

Service Grids: Service grids—as defined by Platform Computing—provide access to resources that can be purchased by corporations to augment their own resources.

4.14 GRID NETWORKS—WILL THERE BE SUCH A THING AS "THE GRIDNET"?

There is an uncanny similarity between the activities taking place in the Grid Computing community—especially in the academic and research realms—and the activities that led to the creation of the Internet. The protocols are the same (IP), work (scientific, R &D) and funding (government, public sector) are similar. The logical question to ask is whether there will be such a thing as the *Grid-Net* or the *Great Global Grid (GGG)* that is independent from the Internet infrastructure.

The commercialization and unprecedented success of the Internet following 1995 led the original users of the Internet—the academic and research community—to move on to other networks. Abilene, Internet2, VBNS, and numerous other networks were built to provide the scientific community their own high speed networks that were not encumbered by commercial traffic.[19] Most of the research community around the world involved in Grid Computing activities today uses these networks.

In August of 2001, the National Science Foundation, an agency of the United States government, funded the Distributed Terascale Facility (a.k.a. Teragrid), which could lead to the potential developing of the *GridNet*. The funding for the Teragrid project also included funding for creating a separate high-speed network that connects participating institutions in Chicago to those in Southern California. It was decided that the current academic and research networks, such as those mentioned earlier, would not be used for the project. There are already plans for additional institutions to join the Teragrid, such as the Pittsburgh Supercomputing Center. We expect that there will be at least a handful of companies wanting to avail the compute and data resources of the Teragrid and will also join the network. If the roster of companies continues to grow, then there is good chance that the Teragrid could be privatized, the same way as the NSFNET was, to support commercial traffic. This could ultimately lead to the creation of the *GridNet* and *GGG*, which operate alongside the Internet.[20]

4.14.1 Grid Network Peering Points

The commercial *Teragrid* or *The GridNet* may still be a few years away, but some of the underlying infrastructure that is required is already being built. One of the key components of the public networks is the availability of peering points. Peering points serve as the point of connection between two or more networks.[21] The exchange of traffic at peering points is through a negotiated peering arrangement between parties and is settlement free.

Star-Light, located at Northwestern University in Chicago, is currently an established peering point for handling research and academic grid networks.[22] The

---

[19] North America: Abilene, Ca*Net 3/4 , DREN, ESNet, MREN, vBNS, vBNS+ Europe: CERN, IsraelIUCC, NorduNet, MIRnet, SURFNet Asia: APAN, CERNET (China), GEMnet (Japan), KOREN, SingAREN, TANet2.

[20] MBone, another parallel network for multicast applications, setup in the late 1990s, however, was not much of a commercial success.

[21] There are two types of peering points. Public peering points are established to facilitate peering between service providers at one place. Private peering points are established bilaterally between two service providers. There are over 150 public peering points around the world that form the basic infrastructure of the Internet. The exact number of private peering points is unknown, but is also expected to be in the hundreds.

[22] For more information please refer to NSF award EIA-9802090 and the Startap Web site *www.startap.net*

funding for Star-Light has been provided by the National Science Foundation, state of Illinois I-Wire program, and University of Illinois at Chicago.

The Star-Light today provides peering at 1 Gbps and 10 Gbps Ethernet speeds. The addition of optical switching will allow Star-Light to offer wavelength-based peering. Currently SurFNet of the Netherlands, CA*net 3/4, I-Wire, and Teragrid are the networks that are peering at Star-Light.

A sister peering point, called the Nether Light, is being established in Amsterdam that will allow research and academic networks to peer in Europe.

It is important to keep an eye on the peering activities occurring at these two sites to determine which institutions are signing on to become part of the grid.

4.15 GRID APPLICATIONS CHARACTERISTICS

The success of Grid Computing will ultimately be determined *not* by the technical sophistication of the Grid Computing protocols, nor by the elegance of grid networks, but by *what* problems Grid Computing will solve.

Not all applications will be able to take advantage of Grid Computing. The grade of parallel efficiency exhibited by an application determines its suitability for Grid Computing deployment. Parallelism is an inherent property of the application.[23] Parallel applications fall under three general categories:

- *Perfect Parallelism*—Also known as *embarrassingly parallel*. An application can be divided into sets of processes that require little or no communication. A Monte Carlo simulation falls into this category.[24]
- *Data Parallelism*—The same operation is performed on many data elements simultaneously. An example would be using multiple processes to search different parts of a database for one specific query.
- *Functional Parallelism*—Often called *control parallelism*. Multiple operations are performed simultaneously, with each operation addressing a particular part of the problem. For example, in a power plant simulation, one application process might simulate the cool system, one the generator, etc. Data communication from one operation to another is required.

---

[23] In case you were wondering, try calculating a Fibonacci series (1,1,2,3,5,8,13,21, . . .) using the formula $F(k+2) = F(k+1) + F(k)$. The result at each step requires the results from the two previous steps, thus these three terms cannot be calculated simultaneously, hence the problem is inherently non-parallelizable.

[24] Monte Carlo simulation; pick random starting point, find solution, save result, repeat until "good enough solution" is found; every processor can do this without communicating with another.

The ratio of the amount of computation done by an application to the amount of communication is called *granularity*. Granularity can be *fine grained, medium grained*, or *coarse grained*. In fine-grained tasks, the operands are small and the communication is high. In coarse-grained tasks, entire programs can be executed on a processor before communication or synchronization is required. Obviously, coarse-grained parallel applications generate the best speedup results.

The amount of data being used by each concurrent task, assuming that the application exhibits parallelism, will determine whether a particular application is suited for being deployed on a desktop grid, high performance grid, or cluster grid. In some cases, desktops may not have the amount of memory required for a particular task or that the network bandwidth requirements may interfere with normal business operations. In each of these cases, applications can then be run on a high-performance grid that is either in-house or available through a utility computing model.

4.16 APPLICATION INTEGRATION

There are numerous parallelization models and programming standards that have been developed and fine tuned over the last decade. An application programmer today can either write a program and define explicit parallelism or use parallelization tools that can analyze an existing program and perform automatic detection of parallelism. Specific compilers allow programmers to parallelize applications simply by adding compiler directives. Table 4.4 illustrates some of the results of applying various techniques to application integration.[25]

The cost and ease of porting applications will be one of the key factors that will determine the success of grid deployments in enterprises. All the leading Grid Computing vendors—such as Platform Computing, Avaki, Entropia, and United Devices—offer various ways to port exiting applications. Entropia, for example, allows binary level integration of applications in its recently released DC5.0 product. Binary level integration allows any native 32 bit Windows application to be integrated without modifying application source code. United Devices provides a software development toolkit (SDK) for its Metaprocessor 3.0 application.

The grade of parallel efficiency exhibited by the applications will determine their suitability for Grid Computing deployment. Parallelism is an inherent property of the application. The greater the ability of the application to be broken down into smaller tasks that can be performed concurrently and independently, the

---

[25] Research performed at NASA Ames Research Center at Moffet Field, California. A computational fluid dynamic application (CFD) was used to benchmark different paralllization techniques.

TABLE 4.4 Parallelizing Applications

Parallization Process	Time/Cost	Performance	Portability	Applicability/ Limitations
Rewritten by hand	Extensive code revision required; error prone	Excellent when implementation tailored to a particular machine	Dependent on portability of standards such as MPI and PVM [1]	Applicable to any code with inherent concurrency
Compiler Directives	Minimal code modification; directives inserted as needed; tuning could be time consuming	Completely dependent on compiler and user's expertise; can be excellent if optimized for caches	Dependent on portability of standards (Open MP)	Performs well for code with simple structure and loop level parallelism
Automated Translation using CAPTools (a FORTRAN code translator)	Tuning needed for good performance	Excellent (after tuning)	Excellent. Library based on standard MPI and PVM libraries	None

[1] Message Passing Interface (MPI) and Parallel Virtual Machine (PVM)

higher the degree of parallelism. If the tasks being performed need to communicate with each other, then the extent of data overhead and communication latency will determine whether the application is in fact getting any benefits from running in a parallel environment.

Assuming that the application exhibits parallelism, then the amount of data being used by each concurrent task will determine whether a particular application is suited for being deployed on a desktop grid, high-performance grid, or cluster grid. In some cases, desktops may not have the amount of memory required for a particular task or that the network bandwidth requirements may interfere with normal business operations. In each of these cases, applications can then be run on a high performance grid that is either in-house or available through a utility computing model.

4.17 GRID COMPUTING AND PUBLIC POLICY

All new technologies bring with them some unintended consequences—some of which are good while others are detrimental to society. In this section we discuss a few of the public policy challenges that may need to be addressed as grid computing proliferates in companies and indeed in consumer homes.

4.17.1 Sleeper Programs

To participate in Internet computing projects such as Seti@Home or Folding@Home, a user downloads a client software package onto their desktop computer. The client is only supposed to perform tasks related to running the particular scientific experiment. But what if the client was actually being used to snoop around the desktop? In this scenario, users would be exposing all the files on their computer through the downloaded client. We have already seen "sleeper" programs being installed on computers when they download certain clients. In one recent example, millions of people that downloaded Kazaa, a popular peer-to-peer file sharing tool were unknowingly downloading a software program that would wake up at a later date and start using the CPU cycles on their desktop. Users should have tools that they can use to set clear policies on what can be accessed on their desktops. Companies should be legally forbidden to clandestinely integrate sleeper programs with clients that are being downloaded by users.

4.17.2 National Security

Certain countries impose strict limitations on high-performance computing exports. Rogue elements, however, with Grid Computing tools can easily create "supercomputers" from simple commodity desktops and use them for nefarious purposes. Although it is easy to regulate export of hardware, it is almost impossible to do the same with software. Even though the Globus Web site specifically states that the toolkit is forbidden for export to "states of concern" such as Iran, Iraq, Libya, and North Korea, we think it will be impossible to monitor and regulate. The question is: Are we unwittingly arming some nefarious characters with a real powerful tool, something that they have been trying hard for many years to attain?

4.17.3 Philanthropic Computing

If a user is volunteering his desktop for a sound, philanthropic cause such as cancer drug discovery, should he not be able to take a tax deduction for the resources he has put forth? The United States Internal Revenue Service allows deductions of $.31 for every mile a car is driven for a charitable cause. We are tempted to petition the IRS for a ruling on how many cents can be deducted for each CPU cycle donated to charitable work. Based on our calculations, at the minimum, a user's electricity bill goes up by 45¢ per day if his computer is utilized at all times.[26]

4.18 CONCLUSION

Grid Computing enables virtual organizations, to share geographically distributed resources. Resources can be supercomputers, clusters, desktop storage systems, sensors, scientific instruments, etc. Grid Computing is not a new concept. It leverages almost a decade's worth of knowledge acquired by high performance computing, cluster computing, peer to peering, and Internet computing communities. Grid Computing protocols are based on protocols developed and refined by the Internet community. Existing protocols have been extended to provide grid-specific functionality. The Open Grid Services Architecture will integrate Web Services functionality into grid protocols. Globus Toolkit is a reference implementation of Grid Computing protocols. It has been a tremendous success and is the most widely deployed in research and academic networks.

The Global Grid Forum is the global standards body for Grid Computing activities. Modeled after IETF, it is providing leadership in setting open grid standards. Grid definitions are based on the *organization* they serve and the *resources* they offer. The numerous marketing names associated with Grid Computing today—such as desktop grids, cluster grids, and enterprise grids—all fall under these two categories. A public Global Grid does not exist yet. But some of the infrastructure that at some point may form the foundation of the *GridNet* is being built today.

---

[26] Assuming idle PC consumes 50W, while a loaded one 200W. Cost per KWH is .125¢.

5 Desktop Grids

David Johnson

Entropia

In This Chapter

- Historical context for Desktop Grid technology
- What is (and what isn't) a Desktop Grid?
- Deployment challenges and value proposition development for Desktop Grid technology
- Key areas to assess when evaluating the suitability of a Desktop Grid for a particular problem or a particular technology environment
- The role of Desktop Grids in an Enterprise computing infrastructure, including the effect of emerging Web Services and Grid Services standards on Desktop Grids
- Examples of ways in which Desktop Grids are being used today to deliver results in a variety of industries

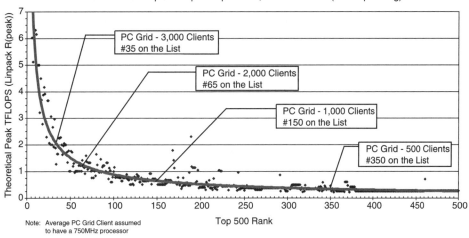

PC Grids Compared with Supercomputer Performance
Source Data: Top 500 Supercomputer List, November 2002 (www.top500.org)

5.1 INTRODUCTION

In this chapter, we will explore the various ways in which a company can reinvent its existing PC infrastructure as an enterprise-class computing resource to provide significant additional capacity for compute-intensive applications without adding significant additional overhead costs.

5.2 BACKGROUND

In this section we cover some of the key lessons learned from the early Internet computing (section 4.6) projects and how these can be applied to creating robust enterprise desktop grids.

5.2.1 Cause Computing and the Internet

Many people are first exposed to the idea of the aggregation of PC processing power through one of the many "cause computing" projects that sprang to life during the last several years. Some examples include searching for extra-terrestrials (SETI@home, *http://setiathome.ssl.berkeley.edu/*), evaluating AIDS drug candidates, (FightAIDS@Home, *http://www.fightaidsathome.org/*), screening for extremely large prime numbers (Greater Internet Mersenne Prime Search, *http://www.mersenne.org/prime.htm*), and predicting climate on a global scale (ClimatePrediction.net, *http://www.climateprediction.net/index.php*). All of these projects are based on the idea of *enrolling*, i.e., a conscious decision on the part of a PC owner to sign up with a particular organization to allow the spare computational cycles of his PC to be used by the selected project. Upon enrollment, a small control program is downloaded to the PC. This program is responsible for communicating with the central project server (using the public Internet connection of the PC) as well as *harvesting* the spare capacity of the machine by executing cause-related computations relayed by the central server. Typically, these projects use relatively short communication packets to drive comparatively long computations on the enrolled PC. This is an attempt to be minimally intrusive on the user and his Internet connection. The method for consuming the spare capacity of the PC can be as simple as executing the cause-related computation in the place of the normal screensaver (taking advantage of those instances when the computer is completely unused) or as complex as executing the cause-related computation continuously as an idle-priority task within the Windows environment (giving preference to any user-initiated tasks and then soaking up any remaining capacity).

These *ad hoc* collections of work-based and home-based PCs from around the world are an example of PC-based *distributed computing* and serve as the forerunners of today's true Desktop Grids. Some of the key concepts that arose out of these projects include:

- Resource Management—All Internet-based grids use passive resource management; they rely on the enrolled PCs to initiate communication with the central administration server on a periodic basis. This limits the degree to which the timeliness of results from such a grid can be predicted. In addition, it limits the ability to reprioritize the computational behavior of the grid (for example, replacing the PC that is working on a particular task) in a timely manner.
- Communication and Data Security—HTTP is the communication protocol between the PCs and the central server. Even if some form of encryption is used in transit, the data usually reside in an unencrypted format on the enrolled PC. This limits the nature of the problems that can be attempted over the public Internet to those in which compromise of the data is not a pressing issue. In addition, in some cases, the answers produced on the enrolled PC may be vulnerable to tampering, causing the confidence in the results to be lower than desired.
- Machine Heterogeneity—A wide variety of machines might be enrolled; these can vary in CPU speed, RAM, hard-drive capacity, and operating system level. The management infrastructure either needs to operate at the lowest common denominator or needs to be aware of differences in the machines and assign tasks appropriately.
- Resource Availability—The entire cause-computing paradigm relies on the idea of voluntary participation. As such, the availability and utility of any particular resource is subject to the whim of the person controlling the PC. The PC may be turned off for the night, the screensaver may be changed, the control program may be disabled (either deliberately or inadvertently), etc. This adds another layer of unpredictability to the performance expectations that can be associated with such a grid.

5.2.2 Distributed Computing in the Enterprise

Distributed computing in the corporate world evolved out of the high-performance computing grids consisting of inter-networked UNIX™ and/or Linux™ machines. Corporate users in that environment had come to expect resource management, security, and availability as an inherent part of their distributed computing infrastructure. Many corporate users realized that the aggregated, unused power of the desktop/laptop PCs assigned to employees represented a large pool of computational cycles that were being wasted and not benefiting their company. However, concerns related to the four concepts discussed above limited the interest and utilization of PC-based distributed computing using the "Internet" paradigm among enterprise users.

In addition, this paradigm did not acknowledge (or attempt to exploit) the fundamental differences between a collection of PCs connected by a corporate intranet and an *ad hoc* assortment of typical home-based PCs connected using the public Internet. These differences include:

- Network Connectivity—Many corporations have their PCs on dedicated high speed (100 Mbps) or very-high speed (1Gbps) networks. A faster, dedicated network connection allows much more freedom in designing a distributed application, as there is not a personal Internet connection to jeopardize. However, some of these machines may have only intermittent or occasional connection to the corporate network (and when they are connected, it may be through a much lower bandwidth pipe). Because many organizations are using portable computers as the primary desktop computing device, both the duration and the quality of any device's connection with the corporate network are difficult to predict.
- Required Participation—Participating in a distributed computing effort can be part of the standard "way of doing things" within the company. This provides more certainty about the composition of the grid, yet does not address any of the robustness issues that remain (PCs may reboot; PCs may be turned off, etc.).
- PC Administration and Security—These are generally already in place so that "sensitive" information can be distributed to the computational nodes without excessive concern. The notion of active management of PCs (for example, the automated "push" of security updates) is an accepted part of corporate PC infrastructure.
- Access to Shared Resources—Most organizations have common data storage on their intranets. This can be as simple as a shared drive mapping that is established in conjunction with a network connection or as complex as a multi-tier, multi-terabyte data warehouse. The Desktop Grid can use this knowledge to reduce or eliminate redundant copies of data and to optimize work assignments based on the knowledge of which members of the Desktop Grid have access to which shared resources.

The technology behind the Internet-based voluntary enrollment methodology for PC-based distributed computing combined with the needs and configuration of enterprise infrastructure to create the idea of a "Desktop Grid" as an analogous concept to a high-performance computing cluster based on UNIX or Linux— something that would allow commodity desktop PCs to be treated as a mission-critical corporate asset and used to do core compute-intensive work vital to the success of the company.

5.3 DESKTOP GRIDS DEFINED

As we learned in the previous section, there might be any number of definitions of a Desktop Grid depending on the assumptions made and options selected. However, the needs of enterprise customers lead us to a specific definition. For the remainder of this chapter, we will consider a Desktop Grid to have the following characteristics:

- A defined (named) collection of machines on a shared network, behind a single firewall, with all machines running the Windows operating system. For simplicity, we will also assume that any single machine is part of one—and only one—Desktop Grid. This named collection may include dedicated machines, intermittently connected machines, and shared machines.
- A set of user-controlled policies describing the way in which each of these machines participates in the grid. These policies should also support automated addition and removal of machines without user or administrative intervention.
- A hub-and-spoke virtual network topology controlled by a dedicated, central server. In other words, the machines on the grid are unaware of each other except as informed by the central server. This makes the Desktop Grid computing model much more of a client-server architecture than a peer-to-peer architecture.
- An actively managed mechanism for distribution, execution, and retrieval of work to and from the grid under control of a central server.

We will also assume that the intent behind the formation of such a Desktop Grid is to aggregate these resources into an easily manageable and usable single (virtual) resource in a fashion that ensures that there is little or no detectable degradation when these computing resources are used for their primary purpose while meeting quality of service, security, and business goals of the larger organization.

Finally, we now introduce a consistent set of terminology for use throughout the rest of this chapter when discussing Desktop Grids, their components, and their uses:

- Grid—This term will be used interchangeably with Desktop Grid for simplicity.
- Grid Server—This is a central machine that controls and administers the Desktop Grid.
- Grid Client—An individual node that is a member of the Desktop Grid from which spare computational resources will be harvested. A Grid Client is typically an existing desktop or laptop PC; however, any Windows-based PC connected to the corporate network can become a Grid Client.
- Grid Client Executive—The software component of the grid infrastructure that resides on a PC, enables that PC to serve as a Grid Client, and manages all interaction between the Grid Client and the Grid Server.
- Work Unit—The packet of computation assigned to a Grid Client by the Grid Server. This packet includes a grid-enabled version of an application, instructions for establishing an environment for the application on the Grid Client, the input data (or a pointer to the location of the input data), and instructions on how to execute the application and produce the output data.

5.4 THE DESKTOP GRID VALUE PROPOSITION

The collection of the existing PCs within an organization typically represents its single largest, untapped computing resource. Average utilization levels for PCs on the corporate desktop range between 5 percent and 8 percent, yet 100 percent of the cost of administration and support for these PCs has already been factored into most corporate accounting schemes. A Desktop Grid solution creates an opportunity to tap into the essentially "free" computing resource represented by these underutilized PCs; this can prove an extremely cost-effective way to increase computing power for most organizations.

A Desktop Grid is sometimes referred to as a "virtual supercomputer." This is not far from the truth. In Figure 5.1, you can see a graph of the top 500 supercomputers in the world as of November 2002 (data obtained from *www.top500.org*).

Even relatively modest grids of a few thousand typical desktop PCs provide computing power that would rank among the fastest 100 supercomputers in the world.

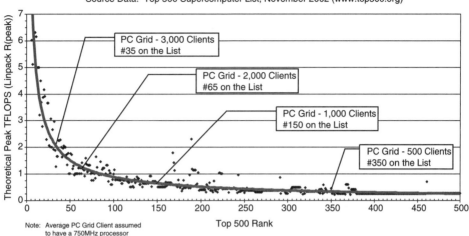

FIGURE 5.1 Relative ranking of various Desktop Grids on the top 500 supercomputer list.

5.5 DESKTOP GRID CHALLENGES

Although harnessing PCs can add horsepower to an organization's overall compute infrastructure, it is not without certain challenges. By its very nature, a grid of PCs provides a compute resource with entirely different characteristics than a static compute platform:

- Intermittent Availability—Unlike a dedicated compute infrastructure, a user may choose to turn off or reboot his PC at any time. In addition, the increasing trend of using a laptop (portable) computer as a desktop replacement means that some PCs may disappear and reappear and may connect from multiple locations over network connections of varying speeds and quality.
- User Expectations—The user of the PC on the corporate desktop views it as a truly "personal" part of his work experience, much like a telephone or a stapler. It is often running many concurrent applications and needs to appear as if it is always and completely available to serve that employee's needs. After a distributed computing component is deployed on an employee's PC, that component will tend to be blamed for every future fault that occurs—at least until the next new component or application is installed.

A loosely coupled network of inherently intermittent computing engines is an extremely hostile environment in which to conduct mission-critical computations.

It is vital that the underlying technology of the Desktop Grid solution is robust in the face of these challenges and, where possible, turns these challenges into advantages.

5.6 DESKTOP GRID TECHNOLOGY—KEY ELEMENTS TO EVALUATE

This section outlines the key elements that need to be considered and incorporated in Desktop Grid technology.

5.6.1 Security

The Desktop Grid must protect the integrity of the distributed computation. Tampering with or disclosure of the application data and program must be prevented. In addition, the Desktop Grid must protect the integrity of the underlying computing resources. The Grid Client Executive must prevent distributed computing applications from accessing or modifying data on the computing resources.

- Application Level—The distributed application should run on the PC in an environment that is completely separate from the PC's normal operating environment. This ensures the security of the PC while distributed application processes are running. The Grid Clients should receive executable programs only from Grid Servers, which should always be authenticated.
- System Level—The Grid Client Executive should prevent an application from using/misusing local or network resources. Machine configuration, applications, and data should be unaffected.
- Task Level—The Grid Client Executive must encrypt the entire work unit to protect the integrity of the application, the input data, and the results.

5.6.2 Unobtrusiveness

The Desktop Grid typically shares computing, storage, and network resources in the corporate IT environment. As a result, usage of these resources should be unobtrusive. The Grid Client should cause no degradation in PC performance, but should utilize every possible computing resource available. When the user's tasks require any resources from the Grid Client, the Grid Client Executive should yield instantly and only resume activity as the resources again become available. The result is maximum utilization of resources with no loss of PC user productivity.

5.6.3 Openness/Ease of Application Integration

The value of a Desktop Grid is directly related to the number, variety, and utility of the applications that can be run on it, combined with the ease with which new or legacy applications can be integrated. A good Desktop Grid system should provide safe and rapid application integration. It should support applications at the binary (executable) level; there should not be any requirement for recompilation, relinking, or access to application source code.

Equally important, the Desktop Grid system must provide application integration in a secure manner, ensuring that any application misbehavior does not adversely affect the desktop user, machine configuration, or network. For example, this might include an inadvertent modal dialog box displayed by the application in response to an error condition. It is not sufficient to just allow the application to execute on the Grid Client; behaviors like this must be buffered within the integrated application environment so that unobtrusive performance is achieved. In addition to application integration security, there are many issues related to overall application suitability for a Desktop Grid; see the section on Desktop Grid suitability later in this chapter.

5.6.4 Robustness

The Desktop Grid must complete computational jobs with minimal failures, masking underlying resource and network failures. Given that PC resources are rarely homogeneous and are often intermittently available (especially true when the PC is a laptop), the grid must execute reliably with fault tolerance on heterogeneous resources that may be turned off or disconnected during execution. Resources contributed by each PC can vary depending on the memory, bandwidth connection, type of processor, and principal use of the system itself. The Grid Server should adapt to these varying resources by matching and dispatching appropriately sized tasks to each machine.

5.6.5 Scalability

With large numbers of PCs deployed in many enterprises, a grid should be capable of scaling to tens of thousands of PCs to take advantage of the increased speed and power a large grid can provide. However, grids should also scale downward, performing well even when the grid is limited in scope.

5.6.6 Central Manageability

The Desktop Grid should provide a central network management capability that allows control of grid resources, application configuration, scheduling, and software version management and upgrades. A single system administrator should be able to manage all of the Grid Clients without requiring physical access to them. Whether the grid comprises 50 or 5,000 PCs, the system must be manageable with no incremental administrative effort as new clients join the system. Management, queuing, and monitoring of the computational work should be easy and intuitive—configurable by priority, access, and typical run configurations.

5.6.7 Key Technology Elements—Checklists

The following checklists summarize the key technology elements to evaluate when considering a Desktop Grid solution:

Security Checklist

- Disallow (or limit) access to network or local resources by the distributed application.
- Encrypt application and data to preserve confidentiality and integrity.
- Ensure that the Grid Client environment (disk contents, memory utilization, registry contents, and other settings) remains unchanged after running the distributed application.
- Prevent local user from interfering with the execution of the distributed application.
- Prevent local user from tampering with or deleting data associated with the distributed application.

Unobtrusiveness Checklist

- Centrally manage unobtrusiveness levels that are changeable based on time-of-day or other factors.
- Ensure that the Grid Client Executive relinquishes client resources automatically.
- Ensure invisibility to local user.
- Prevent distributed application from displaying dialogs or action requests.
- Prevent performance degradation (and total system failure) due to execution of the distributed application.
- Require very little (ideally, zero) interaction with the day-to-day user of the Grid Client.

Application Integration Checklist

- Ability to simulate a standalone environment within the Grid Client.
- Binary-level integration (no recompilation, relinking, or source code access).
- Easy integration (tools, examples, and wizards are provided).

- Integrated security and encryption of sensitive data.
- Support for any native 32-bit Windows application.

Robustness Checklist

- Allocate work to appropriately configured Grid Clients.
- Automatically reallocate work units when Grid Clients are removed from grid either permanently or temporarily.
- Automatically reallocate work units due to other resource or network failures.
- Prevent aberrant applications from completely consuming Grid Client resources (disk, memory, CPU, etc.).
- Provide transparent support for all versions of Windows in the Grid Client population.

Scalability Checklist

- Automatic addition, configuration, and registration of new Grid Clients.
- Compatible with heterogeneous resource population.
- Configurable over multiple geographic locations.

Central Manageability Checklist

- Automated monitoring of all grid resources.
- Central queuing and management of work units for the grid.
- Central policy administration for grid access and utilization.
- Compatibility with existing IT management systems.
- Product installation and upgrade can be accomplished using typical enterprise software management environments (SMS, WinInstall, etc.).
- Remote client deployment and management.

5.6.8 Key Technology Elements—Summary

The platform underlying a Desktop Grid implementation must successfully address all of these technology areas—security, unobtrusiveness, openness, robustness, scalability, and manageability. This provides a necessary, but not sufficient, condition for a successful deployment of Desktop Grids; one that enables a company to harness the untapped power in its existing computing resources with minimal additional risk, leading to accelerated innovation and dramatic increases in the return on investment for an enterprise's computing and network infrastructure. In the next section, we will explore the other critical dimension for success with Desktop Grids, i.e., the degree to which a Desktop Grid is an appropriate tool for solving a particular business problem.

5.7 DESKTOP GRID SUITABILITY—KEY AREAS FOR EXPLORATION

This section discusses the key practical areas that need to be addressed for successful integration of grid computing technology in enterprise desktop environments.

5.7.1 Applications

Even the most technically advanced Desktop Grid deployment is of little use without applications that can execute on it. To be considered as a candidate for execution on a Desktop Grid, you must have a Windows version of the application, all supporting files and environmental settings needed to establish an execution environment for the application, and appropriate licensing to permit multiple, concurrent copies of the application to be executed. Certain applications are better suited for Desktop Grid computing than are other applications; in fact, there is a continuum of "Application Suitability." This section provides information to help you determine where an application resides on this continuum and the degree to which it is suited for distributed execution on a Desktop Grid system.

5.7.1.1 Application Categories

Most applications under consideration for distributed execution fall into one of three categories:

- *Data Parallel*—These applications process large input datasets in a sequential fashion with no application dependencies between or among the records of the dataset.
- *Parameter Sweep*—These applications use an iterative approach to generate a multidimensional series of input values used to evaluate a particular set of output functions.
- *Probabilistic*—These applications process a very large number of trials using randomized inputs (or other *ab initio* processes) to generate input values used to evaluate a particular set of output functions.

5.7.1.2 Analyzing Application Distribution Possibilities

In all cases, it is important to realize that a Desktop Grid system operates by dividing (or distributing) the *input(s)* for a particular application; the application itself is untouched and every work unit uses an identical copy of the application. A careful analysis of the application is required to understand which of its control parameters are hard-coded (and therefore, every work unit must operate using those parameters) and which are changeable based on input values, configuration files, or registry settings (and therefore, each work unit might operate using different values for those parameters). After considering how the application treats its parameters, we have two key considerations, regardless of application category:

- Understanding how to decompose the input(s) of a large, monolithic job into an equivalent set of smaller input(s) that can be processed in a distributed fashion
- Understanding how to recompose the output(s) from these smaller distributed instances of the application into a combined output that is indistinguishable from that which would have been produced by the single large job

We will use the term "*grid-enabled application*" to refer to the combination of an application prepared to execute on a Grid Client, a particular decomposition approach, and a particular recomposition approach. We can now refine our definition of Work Unit to be a package sent to the Grid Client containing the prepared application along with one member of the set of decomposed inputs and instructions for how to create one member of the set of outputs for recomposition.

5.7.1.3 Example–Data Parallel Application

Consider an application that examines a large file of integers (that are stored one integer per line) and counts the number of items greater than a particular target value. The input to this application is the file of numbers and the target value; the output is a single number.

The large file may be split into N smaller files such that each line of the original file is in one (and only one) of the smaller files. The Desktop Grid system processes each of these N files independently using the same target value for the application. Calculate the sum of the outputs from each of the N application instances to recreate the output that would be generated by the large file.

5.7.1.4 Example–Parameter Sweep Application

Consider an application that finds the maximum value of a function F(X,Y) over a specified range using an exhaustive search approach that involves iteration of the parameters. The inputs to this application are start value, end value, and step size for both X and Y. The output of this application is the largest value found for F along with the (X,Y) pair that generated that value.

There are several ways that this application could be grid-enabled. One simple method is to launch a separate instance of the application for each unique value of X that would iterate through the entire range of Y while holding X constant. The output of each of these smaller instances would be an (X,Y) pair along with a value of F. To generate the original output, select the largest value of F among those returned.

5.7.1.5 Example—Probabilistic Application

Consider an application that finds the maximum value of a function F(X,Y) over a specified range using a Monte Carlo approach (generating random (X,Y) pairs within the range and evaluating F). The inputs to this application are a minimum value and a maximum value for both X and Y along with a number of random pairs to be generated. The output of this application is the largest value found for F along with the (X,Y) pair that generated that value.

There are several ways that this application could be grid-enabled. One simple method is to create N instances of the application where each instance generates its proportional share of the total number of random pairs. The output of each of these smaller instances would be an (X,Y) pair along with a value of F. To generate the original output, choose the largest value of F among those returned.

5.7.1.6 Determining Application Suitability

After determining how to decompose and recompose the application inputs and outputs for distributed processing, you can assess where a grid-enabled application falls on the continuum of application suitability. The key measurement to calculate is the Compute Intensity of a typical work unit; this reflects the relative percentage of time spent moving data to and from the Desktop Grid Client compared to the time spent performing calculations on that data. Calculate the Compute Intensity (CI) ratio using this formula:

$$CI = \frac{4 * WorkUnitDuration\ (Seconds)}{InputSize\ (KB) + OutputSize\ (KB)}$$

For example, if we have a grid-enabled application for which a typical work unit executes in 15 minutes (900 seconds) on a hypothetical "average" grid client, consumes 2MB (2,000 KB) of input data, and produces 0.4MB (400 KB) of output data, we can calculate the Compute Intensity to be

$$CI = (4 * 900) / (2000 + 400) = \mathbf{1.5}$$

In general, grid-enabled applications where CI is greater than 1.0 are "well suited" for distributed processing using a Desktop Grid solution. However, this is not a black-and-white decision; if you have particularly efficient data connectivity between your servers and your grid clients (for example, a 1Gb backbone with 100Mb desktop connectivity), then values of CI somewhat less than 1.0 might still yield a well-suited application in your environment. Also, just because an application is not well suited does not mean that it will not work—it means only that the overhead of moving the data back and forth dampens the benefit you will see compared to other applications. One of the best ways to use CI calculations is to help you choose between alternate ways of decomposing data for an application.

5.7.1.7 Fine-Tuning a Grid-Enabled Application

An important factor to include in your evaluation of application suitability is how you plan to receive the benefit of using a Desktop Grid with that application. In general, your benefit will be a combination of these two factors:

- Receiving the same answer faster—By splitting a fixed amount of processing across N work units that can execute in parallel, you will use the power of a Desktop Grid to generate expected results more quickly.
- Receiving a "better" answer in the same time—Using this approach, you hold your expected "time to results" constant and perform significantly more computations during that time. For example, you might explore a parameter space with a finer step size, dramatically increase the number of Monte Carlo trials, etc.

Understanding which of these benefits is more important to you (this will vary by application) will help you choose work unit duration, input/output sizes, parameter sweep step sizes, etc.—all of which have an effect on the CI ratio and make an application more or less suited for execution on a Desktop Grid. Weighing the various tradeoffs is clearly an iterative process and one that is more art than science.

5.7.2 Computing Environment

There are several ways to view the introduction of Desktop Grids to an existing computing environment. In some cases, a Desktop Grid is considered as an alternative, lower-cost option when compared with acquiring new, dedicated computing resources. In other cases, a Desktop Grid is viewed as a complementary addition to an existing infrastructure in which problems with a Windows-based solution can be executed on the Desktop Grid, thereby generating capacity on the other compute devices for problems with only non-Windows solutions. In any case, adding a Desktop Grid to an organization is a substantial change to the overall computing environment. A partial list of environmental considerations includes:

- Archival and Cleanup—Of information on the Grid Server regarding completed computing tasks and their results.
- Backup and Recovery—In the event of failure of a Grid Server (failure of a Grid Client should be automatically handled by the Grid Server).

■ Performance Monitoring and Tuning—Of the Grid Server and the Grid Clients. Is the system operating in an optimal manner based on current usage patterns? Are employees doing their part to ensure maximum availability of the Grid Clients?

5.7.3 Culture

Even after identifying one or more suitable applications for a Desktop Grid, there are a number of cultural considerations that will contribute to the success or failure of a Desktop Grid within any particular organization.

5.7.3.1 At the Desktop–Employee Culture

A well-designed Desktop Grid system will be able to harvest all available computational cycles from the employee's PC without any obvious indication of doing so, provided that the employee leaves the machine in a powered-on state (when not otherwise in use). For many organizations, this requires learning new habits and a different kind of "enrollment"—ensuring that each employee with a PC on the grid understands the vital role his PC is playing in the success of the organization. Some employees may also be very concerned about the Grid Client Executive, perceiving it as a kind of "spy ware" or blaming it for any fault or failure that occurs on their PC. Such attitudes can be overcome with a proactive campaign of employee education that might include informational e-mails, published results (savings in capital expense dollars, faster time to results, etc.), or even a hands-on laboratory where the inner workings of the Grid Server, Client, and Executive may be seen in more detail.

5.7.3.2 In the User Community

Typically, consumers of large quantities of computing cycles are used to a very hands-on interaction between themselves and their computing devices. Current UNIX/Linux cluster systems generally have a fixed, physical presence (i.e., you can actually visit all of the CPUs) as well as a notion of temporary, but exclusive, control (i.e., "I am using the cluster for the next two hours."). Finally, there is an historical prejudice against Windows-based devices with regard to their ability to do "serious" computation.

All of these concerns will diminish over time, provided the users are willing to begin using the Desktop Grid to solve their mission-critical problems. The key here is to identify one or more users with significant unmet computational needs from the current computing infrastructure available to them. An initial reluctance to use a new technology is easy to overcome based on the delivery of initial results—especially if these results could not have been produced in such a timely manner, or with such depth, using the previously available computing resources.

5.8 THE GRID SERVER–ADDITIONAL FUNCTIONALITY TO CONSIDER

We have already identified several essential functions provided by the Grid Server: management/administration of all work units, assignment of work units to Grid Clients, and management/administration of all Grid Clients. The Grid Server can provide a number of other functions that will turn the Desktop Grid from a low-level work unit scheduling system into a fully automated, multifunctional, distributed application execution system. These include:

- Client Group-level Operations—In small (departmental) grids, administering clients on a one-by-one basis is relatively straightforward. As the size and complexity of the grid grows, it is more useful to administer the grid as a collection of virtual, overlapping groups. For example, one group might be all machines located on the second floor, another group might be all Windows XP® machines; these groups will have zero or more machines in common. This notion of Client Groups must be accompanied by a set of rules that allow client membership to be determined automatically for both new Grid Clients and for Grid Clients that have changed status (for example, upgrading the Windows operating system on that client or adding memory to that client).

- Data Caching—The time needed to move data to and from the Grid Client plays an important role in the calculation of Computational Intensity (as described above). Advanced Desktop Grid systems will provide various forms of data caching in which data needed for a work unit can be placed in (or very close to) the Grid Client that will be executing the work unit in advance of the assignment of the work unit to that Client. This caching can either be manually controlled (certain data sets can be pushed to particular Clients and then any work unit that needs those data sets are assigned exclusively to those Clients) or automatically administered (the Grid Server examines its queue of work and ensures that any data needed for a work unit will be available at the Client).

- Job-level Scheduling and Administration—Users will want to interact with the Desktop Grid at a level consistent with their business problems and desired solutions. In general, the Desktop Grid system should support a kind of user interaction substantially similar to "run this application using these inputs with this priority and put the answers here." This is substantially more abstract than the fundamental "work unit" level of the internal workings of the Desktop Grid. The Grid Server should support various levels of job priority along with the ability to select particular Clients (or groups of Clients) for a particular job based on characteristics of the job itself.

- Performance Tuning and Analysis—The Desktop Grid system should provide all necessary data and reports to allow an administrator to determine important performance characteristics of each Grid Client and the grid as a whole. This should include optimum (theoretical) throughput calculations for the grid, actual throughput calculations for any particular job or set of work units, identification of any problematic Clients (or groups of Clients), etc.
- Security—Each separately identified function within the Grid Server user environment should include user-level security—which users may add new applications, which users may submit jobs, which users may review job output, etc. The Grid Server should have its own security system for access to any of its components through direct methods (i.e., any method other than through the supplied user and administrative environments).
- System Interfaces—The Grid Server should support a variety of interfaces for its various user and administrative functions. At minimum, all functionality should be accessible through a browser-based interface. Other interfaces that might be provided include a command-line interface (for scripting support), a Windows-based API (for invoking grid functionality from other Windows programs), and an XML interface (as a general-purpose communication methodology). Any system interfaces provided in addition to the basic browser access must also include a security protocol.

5.9 ROLE OF DESKTOP GRIDS IN AN ENTERPRISE COMPUTING INFRASTRUCTURE

This section discusses the types of grid deployments that can occur within enterprise environments.

5.9.1 Departmental Grids

A very common entry strategy for Desktop Grids within a large organization is a small-scale grid of several tens of PCs under control of a single manager. Grids of this scale are typically too small to warrant serious consideration or attention from an internal IT organization; this allows the initial deployment of a Desktop Grid to concentrate on solving a particular business problem of significance to a particular department without excessive worry about the larger cultural issues discussed above.

5.9.2 Campus Grids

After achieving an initial success on a small scale, a department may advocate the adoption of Desktop Grid technology within the larger organization. Such a rollout would build on the success with a business case that showed that the increase in re-sults significantly outweighs the incremental overhead of adding Grid Clients out-side of the initial department. Once established, a *Campus Grid* becomes a general-purpose distributed computing utility; the initial departments would then be serving as Desktop Grid advocates to other groups within the organization. Without this strong endorsement from the initial user experiences, it is extremely unlikely that just installing a Campus Grid and announcing its availability will lead to a successful outcome—applications and users with unmet computational needs drive adoption.

5.9.3 Web Services and Beyond

The leading edge of thinking regarding distributed computing includes Desktop Grids as part of a universal computing infrastructure created and administered using common industry standards. Today, all commercial Desktop Grid technol-ogy operates as an island of computational capacity with little connectivity to the other computational resources within the organization. This requires users of the Desktop Grid to explicitly "submit" work to the grid and interact with the grid as a standalone resource. Further, there is no ability to provide a true load-sharing fa-cility in which large, computational jobs could be submitted to a central resource administrator and then partitioned among Desktop Grids and UNIX/Linux clus-ters, based on application characteristics and current workload.

The Globus Project (*www.globus.org*) as discussed in section 4.10 is a multi-institutional research effort that seeks to enable the construction of computational grids providing pervasive, dependable, and consistent access to high-performance computational resources, despite the geographical distribution of both resources and users. Computational grid technology is a critical element of future high-performance computing environments that will enable entirely new classes of computation-oriented applications, much as the World Wide Web fostered the development of new classes of information-oriented applications.

The Globus Project assists in planning and building large-scale test beds, both for research and for production use by scientists and engineers. Another important component is the development of the Globus Toolkit, a research prototype con-sisting of a set of robust software tools that runs on a variety of platforms. These tools are being used to develop large-scale grid-enabled applications in collabora-tion with scientists and engineers worldwide.

The Open Grid Services Architecture (OGSA) introduced earlier in section 4.11 is a proposed evolution of the current Globus Toolkit toward a grid system ar-chitecture based on an integration of grid and Web Services concepts and tech-nologies. Described in the Globus-IBM paper, "The Physiology of the Grid," OGSA builds on the widely adopted Globus Toolkit and Web Services technologies. Web Services standards addressed by the paper include Simple Object Access Protocol (SOAP), Web Services Description Language (WSDL), and WS-Inspection.

Existing Desktop Grid solutions will provide support for the Globus Toolkit and OGSA as those standards mature. This will allow Desktop Grids to be integrated seamlessly into an enterprise's computing infrastructure and allow that enterprise to realize the maximum value available from all of its computational resources. Figure 5.2 shows this integrated view.

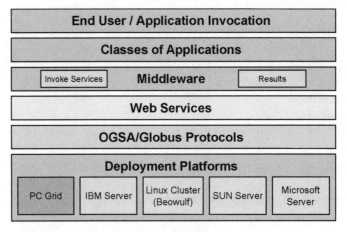

FIGURE 5.2   Web Services and OGSA allow multiple deployment platforms.

5.10 PRACTICAL USES OF DESKTOP GRIDS—REAL-WORLD EXAMPLES

Desktop Grids are gaining acceptance in the corporate world in areas where Windows-based applications are generally available and where many of these applications are particularly well suited to the Desktop Grid paradigm. Some of these key areas include:

- Data Mining—Demographic analysis and legal discovery
- Engineering Design—CAD/CAM and two-dimensional rendering
- Financial Modeling—Portfolio management and risk management
- Geophysical Modeling—Climate prediction and seismic computations

- Graphic Design—Animation and three-dimensional rendering
- Life Sciences—Disease simulation and target identification
- Material Sciences—Physical property prediction and product optimization
- Supply Chain Management—Process optimization and total cost minimization

5.10.1 Example: Risk Management for Financial Derivatives

The following is an example of a grid application from the financial sector.

5.10.1.1 Opportunity

A large North American brokerage organization used a series of interconnected Excel spreadsheets to calculate various risk parameters associated with defined portfolios of securities and fixed-rate derivative instruments. As the desire for more complex evaluations and more scenarios increased, their ability to complete all the required calculations in the time window available (generally overnight) on the fastest PC in their department was in jeopardy. Upper management at this organization also wanted to move the calculation windows from a once per day (batch) paradigm to one in which complex risk-management calculations could be executed on a near-realtime basis.

5.10.1.2 Desktop Grid Solution

This opportunity was ideal for a *Departmental Grid*. All of the PCs in the Risk Management group participated in a small-scale grid that was optimized for these calculations by allowing each PC to have shared access to a central table of risk management parameters. The application was grid-enabled with a combination of parameter sweep and probabilistic techniques without requiring any changes in the core calculations within the spreadsheets. This organization was able to choose a hybrid result (executing "deeper" calculations and receiving the answers faster than before) based on configuration parameters developed during application integration and made available to the job submitter. In this case, a PC grid with less than 25 PCs was more than sufficient to deliver results that were several orders of magnitude improved over that achievable without the Desktop Grid technology.

5.10.2 Example: Molecular Docking for Drug Discovery

The following is an example of a grid application from the life sciences sector.

5.10.2.1 Opportunity

A moderately sized biotechnology servicing company faced capacity constraints as it attempted to respond to its customers' needs for speedy evaluation of new drug candidates against a known database of compounds (called "rapid, flexible protein-ligand docking" within the life sciences industry). One alternative was to purchase additional Linux cluster hardware; this would require capital expense and administrative overhead that were not practical for this organization to take on.

5.10.2.2 Desktop Grid Solution

The required software for the molecular docking existed in a Windows version and this organization had several hundred employees with typical desktop PCs. This presented an ideal application of the data-parallel application integration methodology—splitting the large database of compounds into hundreds of pieces and having each Grid Client evaluate the new drug candidates against a much smaller subset of the larger database. This parallel processing using existing desktop PCs provided throughput comparable to a dedicated Linux cluster that would have cost in excess of $250,000.

5.10.3 Example: Architectural Rendering

The following is an example of a digital rendering grid application.

5.10.3.1 Opportunity

A multinational architectural firm faced a continuing challenge of generating realistic renderings of increasingly complex environments in a timely manner. Their designers all used top-of-the-line PC workstations (for on-demand rendering) and a dedicated rendering farm for batch rendering. They faced several problems with these arrangements: on-demand rendering could tie up a designer's PC for long periods, scheduling on the dedicated rendering farm was strictly FIFO (first in, first out) with little opportunity to control rendering once submitted to the farm, and there were vast pools of computational capacity going to waste during off-hours.

5.10.3.2 Desktop Grid Solution

The basic notion of distributing the rendering of a single frame among multiple rendering engines was already an inherent part of the rendering software used by this firm; their existing batch-rendering farm already used this approach. Their problems stemmed more from the inefficient use of available resources; for example, there were idle designer workstations at their European and Asian locations during peak working hours at their North American location. This organization gained tremendous advantage by turning every designer workstation and each of the dedicated rendering PCs into a Grid Client. This allowed for central management of all of these high-end computing resources as well as using the Grid Client

Executive to manage the individual designer workstations. There were some necessary cultural changes in conjunction with the Desktop Grid implementation associated with a perceived "loss of control" from the designers who were discouraged from using their own workstation for on-demand rendering. This concern evaporated quickly once the designers saw the dramatic increase in overall throughput available with the Desktop Grid.

5.11 CONCLUSION

Desktop Grids started just a few years ago as noble-minded projects for combining spare compute capacity of individual PCs to serve a common cause; they have turned into a technology that can aggregate the unused cycles of an organization's existing PC resources into a powerful, virtual computing engine. This technology can be used to supplement (or even replace) existing high-performance computing resources at a fraction of the cost, as it uses existing hardware that is already fully amortized and supported. As we have seen, not all computing problems are well suited for resolution with a Desktop Grid; however, for those that are, it can provide a new, viable alternative and generate spare capacity on scarce server-based resources.

6 Cluster Grids

Ian Lumb

Platform Computing, Inc.

In This Chapter

- Clusters Defined
- Industry Examples of Clusters
- Cluster Grids

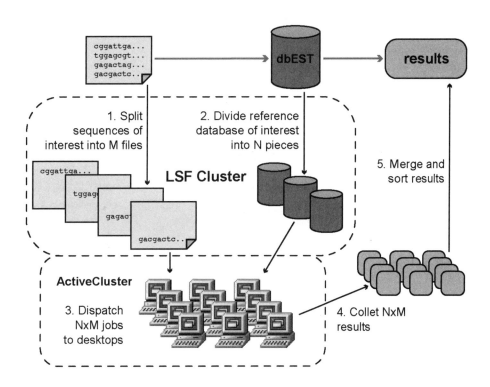

6.1 INTRODUCTION

"*Cluster*" is an overloaded term. In a classic reference on clusters, Gregory Pfister defines a cluster as ". . . a type of parallel or distributed system that consists of a collection of interconnected whole computers, and is used as a single, unified computing resource." In this chapter, the following definition is used: "A cluster is a logical arrangement of independent entities that collectively provide a service."[1]

This definition is consistent with Pfister's, but covers a broader range of usage:

- "Logical arrangement" implies a structured organization. *Logical* emphasizes that this organization is not necessarily static. Smart software and/or hardware is typically involved.
- "Independent entities" implies a level of distinction and function outside of a cluster context that may involve a system or some fraction of a system (e.g., an operating system).
- "Provide a service" implies the intended purpose of the cluster. Elements of pre-service preparation (i.e., provisioning), and post-service teardown, may be involved here.

Armed with this definition, it is possible to explore the cluster landscape. Although clusters are the subject of numerous articles, books, discussion forums, etc., there is no framework for the cluster landscape. Figure 6.1 introduces a framework via a plot of the virtualized instance as a function of the achieved external appearance.

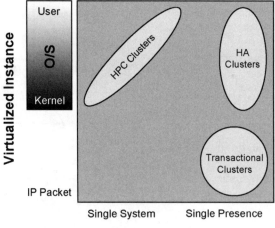

FIGURE 6.1 Cluster landscape as revealed by virtualized instance as a function of external appearance.

---

[1] Pfister, G. F., *In Search of Clusters*, Second Edition, Prentice Hall, 1998.

The vertical axis represents the "independent entity." To cover the cluster landscape, a range of instances from IP packets to the operating system—both user and kernel-space components—are identified. To collectively provide a service, it is these instances that are virtualized. The horizontal axis represents the "logical arrangement" of the distinct entities. Although the external appearance is always of a single service, the specifics vary from a single system to a single presence. Three distinct features emerge on this two-dimensional landscape: Clustering for the purpose of High Performance Computing (HPC, upper left in Figure 6.1), High Availability (HA, upper right in Figure 6.1), and Transactional Services (lower right in Figure 6.1). The technology (6.2 Clusters) and real-world use (6.3 Industry Examples) of HPC clusters is the focus of this chapter. Placing clusters in the context of Grid Computing is considered toward the end of the chapter (6.4), which closes with a summary (6.5).

6.2 CLUSTERS

In the past, Symmetric MultiProcessor (SMP) and vector architectures competed for dominance in the HPC space. Today, 342 of the Top 500 Supercomputer Sites use cluster-based architectures.[2] HPC clusters (upper left in Figure 6.1) use Smart System Software (SSS) to virtualize independent operating-system instances to provide an HPC service. Next to the attractive price/performance of COTS components, SSS plays a key role here. SSS allows a number of distinct systems to appear as one—even though each runs its own instance of the operating system. As Figure 6.2 illustrates, there are two possibilities for SSS. At one extreme the Single System Image (SSI) is SSS that involves *kernel modification*. At the other extreme, the Single System Environment (SSE) is SSS that runs in user space as a layered service. The arrows in Figure 6.2 emphasize interconnections and corresponding

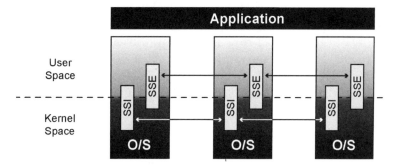

FIGURE 6.2 A layered view of SSS opposite the operating system and end-user applications.

[2] Top 500 Supercomputer Sites, *http://www.top500.org*

communications. Though presented as extremes, examples of SSI-SSE integration do exist. To varying degrees, these solutions enable computing for capacity (i.e., throughput of serial, parametric, and embarrassingly parallel applications) and/or capability (i.e., multithreaded and distributed-memory parallel applications).

6.2.1 Single System Image

With the Cray T3E[3] as archetype, it is clear that clustering through SSI is not "new." Recent emphasis centers on the Linux operating environment, which continues to drive the commoditization of HPC. Although there are a number of SSI solutions available, Beowulf clustering is representative of the approach.

In 1994, Thomas Sterling and Donald Becker of CEDIS[4] created the first Beowulf[5] cluster out of 16 Intel 486 generation systems interconnected via channel-bonded Ethernet. An instant success at the time, Beowulf clustering has reached beyond academic circles in its ability to generate attention, and has become an acknowledged genre in HPC.

Beowulf clusters consist of:

- Commodity-off-the-shelf (COTS) hardware
- LAN interconnect technology
- The GNU/Linux operating system
- Smart system software
- Programming environments

The enabling aspect of Beowulf clustering derives from the SSS that allows interconnected COTS-class hardware, each running its own instance of GNU/Linux, to function as a distributed-memory parallel compute engine.

Individual GNU/Linux kernel instances abstract process manipulation through the notion of a process identifier (PID). However, no such abstraction exists for the case in which a parent process has forked child processes that execute on physically distinct systems—each running their own instance of the GNU/Linux kernel. In other words, there is no concept of a *global PID* for clustered Linux systems. To address this shortcoming, the Beowulf SSS incorporates a kernel modification to provide a *distributed process space* (BPROC). BPROC allows:

---

[3] Cray T3E, *http://www.cray.com/products/systems/t3e*

[4] The Center of Excellence in Space Data and Information Sciences (CESDIS) is a division of the University Space Research Association (USRA) located at the Goddard Space Flight Center in Greenbelt, Maryland. CESDIS is a NASA contractor, supported in part by the Earth and Space Sciences (ESS) project. The ESS project is a research project within the High Performance Computing and Communications (HPCC) program.

[5] Beowulf.org, *http://www.beowulf.org*

■ PIDs to span multiple physical systems—each running their own instance of GNU/Linux

■ Processes to be launched on multiple physical systems—each running their own instance of GNU/Linux

This distributed process space is key to the creation of a clustered Linux environment for distributed-memory parallel computation.

Ultimately, the hardware, interconnect technology, operating system, and SSS collectively provide an infrastructure for scientists and engineers to develop and execute applications. Beowulf clusters support distributed-memory parallel computing via the Parallel Virtual Machine (PVM)[6] or via the Message Passing Interface (MPI).[7] Complimentary GNU compilers and tools (e.g., editors, debuggers, profilers, etc.) are also available.

First-generation Beowulf clusters generated interest for a variety of reasons:

■ They demonstrated that a substantial parallel computation infrastructure could be built from readily available, and inexpensive, hardware and software components.

■ They leveraged key, existing open source software in the GNU/Linux operating system, plus the programming environments offered by PVM and MPI in tandem with the GNU compilers and tools.

■ They demonstrated that a distributed-memory parallel computation approach could rival, and in some cases surpass, the performance characteristics of more-traditional serial or shared-memory programming paradigms.

In short, Beowulf clustering placed distributed-memory parallel computation in the public domain by making it simultaneously accessible and realizable.

Touted as the next-generation solution, the Scyld[8] Beowulf commercial distribution offers the following enhancements:

■ Installation and administration improvements
■ Efficient, single-point distributed process management
■ Various 64-bit capabilities
■ Distributed-process-space-aware MPICH[9]
■ MPI-enabled linear algebra libraries and Beowulf application examples

---

[6] PVM, *http://www.epm.ornl.gov/pvm*

[7] MPI, *http://www.mpi-forum.org/*

[8] Scyld, *http://www.scyld.com*

[9] MPICH is Argonne National Laboratory's open source implementation of MPI. It is widely used in Linux and other operating environments. MPICH, *http://www-unix.mcs.anl.gov/mpi/mpich*

The second-generation Beowulf clustering solution provides a number of significant enhancements beyond what was available in the first-generation solution. Despite this progress, Sterling and colleague D. Savarese identify the software barrier:

> *The earliest Beowulf-class systems were employed as single-user systems dedicated to one application at a time, usually in a scientific/engineering computing environment. But the future of Beowulf will be severely limited if it is constrained to this tiny niche.*
>
> *The need to enhance Beowulf systems usability while incorporating more complicated node structures will call for a new generation of software technology to manage Beowulf resources and facilitate systems programming.[10]*

In terms of resource management, Sterling and Savarese elaborate:

> *Adequate system management may depend on the virtualization of all its resources. This will separate the user application processes from the physical nodes upon which the tasks are executed. The result is a system that dynamically adapts to workload demand, and applications that can perform on a wide range of system configurations trading time for space. Therefore, a new class of workload scheduler will be required, developed, and incorporated in most Beowulf systems. It will support multiple jobs simultaneously, allocating resources on a to-be-defined priority basis. It will also distribute the parallel tasks of a given job across the allocated resources for performance through parallel execution. Such schedulers are not widely available on Beowulfs now and will be essential in the future. They will incorporate advanced checkpoint and restarting capabilities for greater reliability and job swapping in the presence of higher priority workloads. Compilers to use the more complicated structures of the SMP nodes will be required as well to exploit thread level parallelism across the local shared memory processors. The software used on these systems will have to be generally available and achieve the status of de facto standard for portability of codes among Beowulf-class systems.[10]*

Sterling still regards software as the key barrier, as Beowulf clusters move into their third plateau: "The obstacle to this next quantum step is in enabling software. There are three main challenges to next generation commodity clusters that may limit their long-term impact: resource management, fault recovery, and programming methodology."[10]

In part, Sterling calls for a shift toward an even purer form of SSI. For example, this shift would require global namespaces for entities in addition to BPROC's

---

[10] Savarese, D. F. & T. Sterling, Beowulf, in R. Buyya (ed.), *High Performance Cluster Computing*, Volume 1, 625-645, 1999.

global PIDs, and impact directly on the three challenge areas. Although CPLANT,[11] MOSIX,[12] and Unlimited Linux have taken encouraging steps in this direction, there remains a gap in achieving SSI in the sense of the Cray T3E.

Outside of resource-management, it is noted that:

- There is a tight dependency between Beowulf's BPROC and the Linux kernel (Figure 6.2). Even in the case of routine upgrades and/or patch application, this dependency needs to be considered, and implies system downtime. The Two-Kernel Monte, use of diskless compute nodes, or Linux BIOS,[13] each reduce this strong kernel interdependency.
- The architecture of BPROC itself has caused scalability concerns. Enhanced scalability is a current focus in BPROC development.
- As a Linux-kernel modification, licensing for BPROC is under the GNU Public License (GPL). The viral nature of the GPL makes it challenging for commercial independent software vendors (ISVs), Linux distribution providers and integrators, plus traditional-UNIX system vendors, to simultaneously add value and retain a differentiating edge.

6.2.2 Single System Environment

Clustering solutions can also be delivered via an SSE. In contrast to SSI, clustering via SSE does not require modifications to the kernel (Figure 6.2). Instead, SSE runs in user space and provides a distributed process abstraction that includes primitives for process creation and process control.

The user-space approach releases the single-operating-system restriction, and allows third parties to craft cross-platform clusters based on Linux, Mac OS, UNIX, and/or Windows. SSE directly addresses the tension between supply and demand by matching an application's resource requirements with the resources capable of filling the need. By effectively arbitrating the supply-demand budget over an enterprise-scale IT infrastructure, subject to policy-driven objectives, SSE solutions allow organizations to derive maximal utilization from all available computer resources.

In general, SSE solutions make use of dynamic-load-state data to assist in making effective, policy-based scheduling decisions, and in applying utilization rules to hosts, users, jobs, queues, etc., all in real time. This dynamic-load-state capability has significant implications, as task-placement advice is provided directly to the application on dispatch for execution.

---

[11] CPLANT, *http://www.ca.sandia.gov/cacplant*

[12] MOSIX, *http://www.mosix.org*

[13] Linux BIOS, *http://www.linuxbios.org/*

A remote-execution service is required to allow computational tasks to be communicated over a network. Although the standard remote-shell infrastructure (i.e., rsh-rshd-rexec) is a possibility, a more-comprehensive service is required to allow for:

- Authenticated communications over a network
- A high degree of transparency in maintaining the user's execution environment
- Task control with respect to limits, signal passing, etc.

Although SSE solutions do not need to provide a distributed-process space, task-tracking mechanisms are required. Thus application task identifiers act as a handle to the individual (parent and child) processes that collectively comprise a distributed application. In addition to providing a unique identifier for application control, such cluster-wide identifiers can be used in monitoring, manipulating, reporting, and accounting contexts.

SSE solutions typically employ a policy center to manage all resources—e.g., jobs, hosts, users, queues, external events, etc. Through the use of a scheduler—and subject to predefined policies—demands for resources are mapped against the supply for the same in order to facilitate specific activities.

The extension of SSE solutions to support the programming, testing, and execution of parallel applications in production environments, requires:

- Complete control of the distributed processes comprising a job to ensure that no processes will become unmanaged. This effectively reduces the possibility of one parallel job causing severe disruption to an organization's entire compute infrastructure.
- Vendor-neutral and vendor-specific MPI interfaces.
- The ability to leverage a policy-driven SSE infrastructure, that is cognizant of dynamic load state.

Challenges specific to the management of MPI parallel applications include the need to:

- Maintain the communication connection map
- Monitor and forward control signals
- Receive requests to add, delete, start, and connect tasks
- Monitor resource usage while the user application is running
- Enforce task-level resource limits
- Collect resource usage information and exit status upon termination
- Handle standard I/O

To illustrate the value of parallel-application management for developers of MPI applications, consider the following example of fault tolerance. It is beyond the present scope of the MPI, and indeed PVM, to take into account transient situations (e.g., a host that exhibits a kernel panic and crashes) that inevitably occur while an application is in the execution phase. Such situations will affect *some* of the processes involved in the execution of the MPI-based application. If the application does not include some mechanism to address such situations, it is possible for the remainder of the application to run to completion, and deliver incomplete and (potentially) meaningless results. The effect of this situation is compounded when attempts are made to interpret the results. SSE solutions cannot enable fault tolerance in MPI-based applications to the degree that resynchronizations and reconnections are made possible. However, SSE can trap and propagate signals, thereby improving on the degree of management during execution—all without the need for additional coding (beyond exception handlers) by the MPI application developer.

6.3 INDUSTRY EXAMPLES

SSS is an increasingly strategic component of the overall HPC infrastructure in a number of contexts; three application areas can be identified:

- Capacity HPC. Typified by serial-processing requirements, capacity HPC is highly focused on throughput. Often, individual tasks are loosely related, and a parametric processing approach can be applied. Compute farms based on SSE are of common use in this case.
- Capability HPC. Typified by parallel processing requirements, capability HPC is focused more on the resources required to address challenging problems than on throughput requirements. Traditional Grand Challenge problems fall into this category. Solutions based on either SSI or SSE are common in this case.
- Hybrid HPC. Capacity and capability tend to be idealized end points. Real-world examples involve some combination of capacity and capability requirements. SSE handles the mixed workloads found in this case.

In the remainder of this section, the intention is to provide specific, real-world examples drawn from each of the above classifications. Our examples are drawn from electronic design automation (EDA), the life sciences, and industrial manufacturing.

6.3.1 Electronic Design Automation (EDA)

The high-tech field of electronic design automation (EDA) offers rich possibilities for illustrating SSE in capacity-driven simulation (Figure 6.3). In EDA, the fundamental challenge stems from incremental progress into deeper sub-micron design technologies; this advance implies staggering challenges for design synthesis, verification, timing closure, and power consumption. Through direct association with Moore's Law, design synthesis has gained a profile. However, it is design verification that has an even greater potential to become the ultimate design bottleneck: As design complexity increases, verification requirements escalate rapidly.

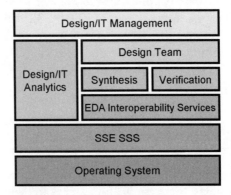

FIGURE 6.3   SSE enhanced productivity stack for electronic design automation (EDA).

Synopsys[14], one of the industry leading EDA tool vendors, proposed a new paradigm to address the ever-increasing verification resource bottleneck: Use fast functional verification at the register-transfer level (RTL), and static verification at the gate level. Although Synopsys offers an end-to-end solution consistent with this new verification paradigm, attention here focuses on functional verification where use is made of Synopsys VCS. This tool translates Verilog[15] source code into a highly optimized executable simulation through the use of a C or its own native compiler. Thus the upshot of functional verification with Synopsys VCS, in terms of simulation workload, is a number of jobs that together form verification/regression suites—e.g.:

---

[14] Synopsys, *http://www.synopsys.com*

[15] Verilog is a hardware design language (HDL) commonly used in the design of, e.g., application-specific integrated circuits (ASICs).

```
vcs <options> test_vector1
vcs <options> test_vector2
. . .
vcs <options> test_vectorN
```

where N tests are run through sequentially. Each Synopsys VCS simulation run can be easily cast, submitted, executed, and summarized as a series of jobs for Platform LSF.[16] Alternatively, this sequence, or array of jobs can be logically grouped together as a job array for parametric processing via Platform LSF. Individual simulations can run for a few minutes or for many hours.

Since 1992, Nortel Networks has used Platform LSF in numerous groups. For example, in Nortel's Optical Networking Group (Ottawa, Ontario, Canada), an Application Specific Integrated Circuit (ASIC) compute farm is comprised of various uni-to-dual-processor Sun Ultra 10/60/80 systems plus Sun Enterprise 420/450/3500/6500 SMPs; collectively these Solaris-based systems amass 150 CPUs. Through their implementation of Platform LSF, the primary benefits derived by this group can be itemized as follows:

- Increased ability to handle high-volume ASIC regression requirements. Literally hundreds of regression jobs can be submitted, monitored, and managed by a single regression operator using Platform LSF.
- Compute capacity is automatically allocated where it is needed the most. Simulation and timing analyses are memory- and CPU-intensive. Platform LSF facilitates the request for these demanding resource requirements by allocating jobs of this type to the medium-to-large Enterprise class SMP servers.
- Actual compute capacity requirements are tracked historically via Platform Intelligence—an IT analytics solution. This historical data is used proactively:
  - By design operators to better specify resource requirements on a per-job basis.
  - By LSF administrators to better tune queue configurations for optimal usage—especially for jobs on a product's critical path. In practice, it has been validated that attempting to oversubscribe resources can introduce delivery delays, as much as designers having to wait for resources to be released.
  - To perform ongoing resource needs analyses. Such efforts ensure that projects remain on schedule by introducing additional capacity as required.
- Effective partitioning of the varied workloads. While regressions have leveraged the 60 CPUs of designers' workstations, especially during off-peak hours, resource-hungry simulations and timing analyses have been able to make use of medium-to-large SMPs within the farm.

---

[16] Platform Computing, Inc., *http://www.platform.com*

■ Virtual engineers work round the clock—i.e., 24/7/365. By way of a robust infrastructure that minimizes the effects of transient situations (e.g., software licenses being unavailable), ASIC developers are able to achieve quantifiable
gains in productivity.

Raw output from the bacct -l <login> command of Platform LSF provides a
solid example to support the statements made above:

```
SUMMARY:           ( time unit: second )
Total number of done jobs:        4520
Total number of exited jobs:       666
Total CPU time consumed:     0370296.0
Average CPU time consumed:      3927.9
Maximum CPU time of a job:     62378.6
Minimum CPU time of a job:         0.0
Total wait time in queues: 18674844.0
Average wait time in queue:     3601.0
Max. wait time in queue:      171316.0
Minimum wait time in queue:        0.0
Avg. turnaround time (sec/job): 10235
Maximum turnaround time:        432675
Minimum turnaround time:             6
Average hog factor of a job:      0.31
( cpu time / turnaround time )
Maximum hog factor of a job:      1.00
Minimum hog factor of a job:      0.00
Total throughput:                 2.41 (jobs/hour) during 2153.20 hours
Beginning time:         Sep 18 16:21
Ending time:            Dec 17 08:33
```

This data, which pertains to Synopsys VCS jobs, was collected over a three-
month period. From this data, the average turnaround time for successfully
completed jobs, namely 10,235 seconds/job/CPU, normalized to hours, is approximately 2.84 hours/job/CPU.

For a designer who needs to run a regression suite consisting of 200 individual
Synopsys VCS jobs, the implied total turnaround time is almost 12 days on a dual-
processor Sun Enterprise 450 SMP system—(2.84 hours/job/CPU * 200 VCS jobs /
2 CPUs). In striking contrast, a 50-CPU workstation cluster can turn this workload
around overnight (i.e., in less than 12 hours!)—(2.84 hours/job/CPU * 200 VCS
jobs / 50 CPUs).

Clearly, compute farms that incorporate compute capacity from idle workstations can deliver impressive throughput performance. It is important to reiterate that this approach also renders the dual-processor Sun Enterprise 450, with its more extensive memory capacity, available for more-appropriate simulation and timing analyses. The regression-suite productivity enhancement is compelling, as more than a 95 percent improvement was achieved—without the need to introduce additional hardware. Clearly, SSE offers material benefits in addressing the verification bottleneck.

6.3.2 Bioinformatics

From the perspective of workload management via SSE, the bioinformatics aspect of life sciences has much in common with EDA. In fact, a search-and-replace on "Synopsys VCS" with "NCBI BLAST"[17] in the previous example would allow for a representative illustration of capacity HPC in the case of genome sequence analysis (GSA). The example of GSA also permits the introduction of Desktop Clusters. Platform ActiveCluster shares much in common with the Desktop Grid solutions described elsewhere (Chapter 5). However, as illustrated in Figure 6.4, Platform ActiveCluster can be

---

[17] NCBI BLAST, *http://www.ncbi.nlm.nih.gov/blast*

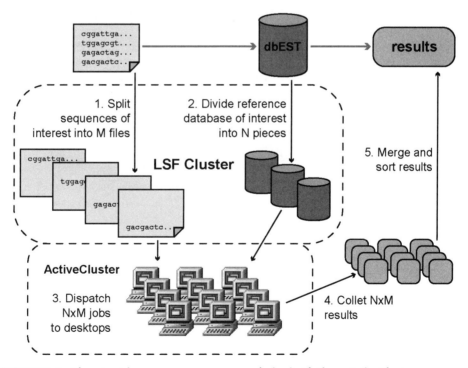

FIGURE 6.4 Five steps in genome sequence analysis via Platform ActiveCluster.

cast as a complimentary addition to a cluster based on Platform LSF. In this case, desktops pull workload for processing on an opportunistic basis. This contrasts with the proactive push to the dedicated resources managed by Platform LSF. Figure 6.4 also provides a visual representation of a standard GSA use case scenario. In five steps, it's made clear that the embarrassingly parallel nature of GSA can be fully exploited. Thus the opportunistic and dedicated HPC resources, virtualized by Platform ActiveCluster and Platform LSF, respectively, collectively form an HPC cluster.

6.3.3 Industrial Manufacturing

Computed-assisted engineering (CAE) continues to play an increasingly significant role in industrial manufacturing (Figure 6.5). As a consequence, globally distributed teams are enabled to collaborate on design, crash, and structural analyses, plus understanding the dynamics of fluid flow. Given that engineer's tools provide reasonably accurate representations of real-world engineering situations, the primary challenges for the IT infrastructure relate more to obtaining high-fidelity results in a cost-effective fashion.

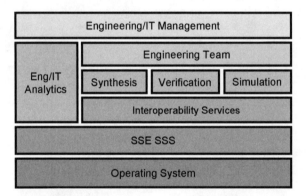

FIGURE 6.5 SSS enhanced productivity stack for industrial engineering.

Computational fluid dynamics (CFD) in an automotive manufacturing context provides a particularly compelling illustration of what can be achieved. Fluent, Inc.[18] is an independent software vendor (ISV) that provides software technology for fluid-flow simulations in three-dimensions. For some time, Fluent has employed a domain-decomposition strategy in their back-end solver. Particularly well suited for parallel processing, current customer deployment typically involves an MPI implementation of the solver across a cluster of interconnected workstations.

---

[18] Fluent, *http://www.fluent.com*

Ford Motor Company's enterprise VISTEON implemented an MPI version of Fluent across a cluster of Hewlett-Packard workstations.[19] The goal was to provide a compute infrastructure that could simulate complex airflows and heat-transfer phenomena in automotive heating, ventilation, and air conditioning (HVAC) units. At implementation time, simulations were comprised of 500,000 elements, with the expectation that this number could only increase.

Platform HPC was implemented to manage this capability processing. Almost completely transparent to the engineer, Platform HPC facilitated the allocation of all compute resources between all classes of application workload—not just the CFD simulations. The ability to schedule jobs requiring multiple processors on physically distinct systems, and to ensure that this processing is completed reliably, were identified as key benefits. Because engineers' workstations were used interactively during regular working hours, Platform HPC was also required to checkpoint executing simulations at the start of each workday and restart them at the end. In this case, Platform HPC was tightly integrated with Fluent to provide highly efficient checkpoint/restart capability at the application level. This same checkpoint/restart capability can be used to provide job migration—e.g., in the case when a host experiences a system crash.

This same integration provides a number of additional seamless benefits:

- Because Platform HPC is cognizant of the dynamic load state across the workstation cluster in real time, it provides Fluent with a list of least-loaded hosts on which the simulation can be run. Thus the dynamic scheduling of the MPI-based Fluent solver can leverage an optimal array of distributed processors. In contrast, MPI applications are typically scheduled with information provided through a host's file that needs to be updated manually.
- The rich workload policy infrastructure of Platform HPC can be used. In addition to a fairshare policy, policies specific to parallel computation can be applied—e.g., advance reservation, backfill scheduling, memory or processor reservation, etc. The open-architecture, modular nature of Platform HPC also allows for custom policies. A scheduling API accelerates the development of these custom policies.
- Controlling and reporting on applications that have been distributed across a network can be challenging. Platform HPC provides a comprehensive infrastructure that allows for signal propagation, limit enforcement, plus realtime resource consumption data.

Commercial ISVs have shown a tremendous willingness to deliver solutions based on parallel processing via the MPI in industrial manufacturing. When used in tandem with a resource management infrastructure such as Platform HPC, this capability processing can provide for high-quality results in a timely fashion.

---

[19] FLUENT on Hewlett-Packard Linux, HP-UX, and Windows NT Clusters, Technical white paper, December 1999.

6.4 CLUSTER GRIDS

Based on their success with clusters, customers in EDA and industrial manufacturing sought to take virtualization to the next level. The resulting federation of clusters had to:

- Allow for selective sharing of resources
- Preserve a high degree of local autonomy
- Address availability on a cluster-by-cluster basis

Fundamental to each of these requirements is the need to span geographic boundaries within an organization.

Platform MultiCluster was introduced in 1996 to federate clusters based on Platform LSF. Platform MultiCluster addresses the identified requirements, and is deployed in one of the three submission-execution topologies (Figure 6.6). Also shown in Figure 6.6 is the single-submission, single-execution (SSSE) case—i.e., a single cluster. Based on this practical experience, it's clear that an HPC cluster is not a grid. A grid needs to involve more than one cluster, address collaborative requirements, and span some geographic distance.

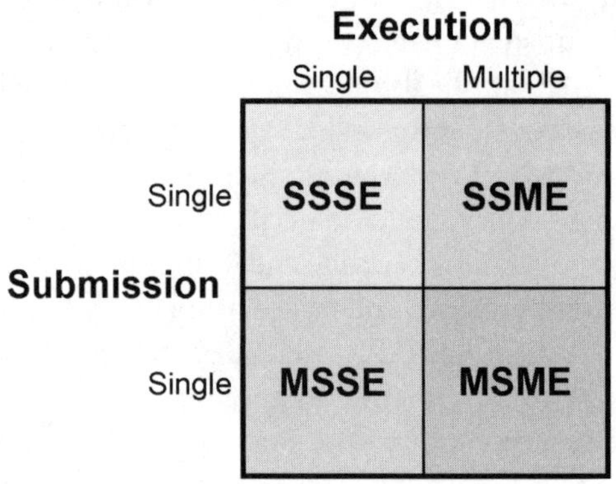

Execution

	Single	Multiple
Single	SSSE	SSME
Single	MSSE	MSME

(Submission)

FIGURE 6.6 Submission-execution topologies for Platform MultiCluster.

To better define the term, Ian Foster has introduced a three-point grid checklist: ". . . a grid is a system that (1) coordinates resources that are not subject to centralized control using (2) standard, open, general-purpose protocols and interfaces to (3) deliver nontrivial qualities of service."[20] Each of these points requires elaboration:

- Coordinates resources that are not subject to centralized control. Because geographic distribution of people and resources is common, coordination between multiple departments within a single organization or multiple organizations may be necessary. In some cases, cooperation among organizations exists only for a finite period of time, so the term *virtual organization* is often used. Trust relationships and connectivity are examples of concerns between cooperating parties.
- Uses standard, open, general-purpose protocols and interfaces. An increasingly real vision at the present time, the emerging Open Grid Services Architecture (see Chapter 9) holds the promise for refactoring existing and developing new technology around open standards.
- Delivers nontrivial qualities of service (QoS). Often these qualities of service are referred to as Service Level Agreements (SLAs) or policies. In the Grid Computing context, QoS translates business objectives into objectives for the IT infrastructure, thus enabling effective utilization, resource aggregation, and remote access to specialized resources.

In his checklist, Foster is adamant that clusters are not grids. Even though clusters may have grid-like affinities, they are:

- Centrally controlled, but not distributed geographically. In the case of HPC clusters, Beowulf's BPROC and the master of a Platform LSF cluster, are these points of central control. Preferring to expose only a few nodes as access points, clusters often make inaccessible those nodes dedicated for compute purposes. This architecture, under the control of a single organization, greatly simplifies the trust relationships that need to be addressed. Even in the case of compute farms, clusters tend to be tightly coupled architectures. Although this tight coupling permits use of shared file systems, it is not tolerant of the latencies that would inevitably result from the introduction of significant geographic distance.
- Often built around proprietary technologies. In the case of clustering for HPC via SSE, proprietary protocols and interfaces remain in use today. The emerging OGSA is expected to be of influence here. Use of proprietary software and hardware technology is even more pronounced in the case of clustering for High Availability and transactional processing.

---

[20] Foster, I., "What Is the Grid? A Three Point Checklist," *GRIDtoday*, **1**(6), July 22, 2002. Available online at *http://www.gridtoday.com*

■ Delivering non-trivial QoS. Not only a significant motivation for the use of a cluster solution, non-trivial QoS is a very strong point of alignment between clusters and grids. What differentiates clusters from grids on QoS is the locality of concern—a tightly coupled LAN versus a geographically dispersed WAN environment. Instead of non-trivial QoS, or even SLAs, the term policy tends to be used in the case of clusters.

All of the above suggests a slight modification to the cluster definition introduced at the outset:

"A cluster is a local-area, logical arrangement of independent entities that collectively provide a service."

By introducing "tightly coupled," the anticipated latency introduced by geography frames the definition. Pfister's definition is not explicit on this point. It is important to reiterate that the degree of coupling varies within clusters. For example, clusters designed for MPI parallel computing via low-latency, high-bandwidth interconnects are certainly coupled more tightly than are compute farms used in high-throughput situations. Thus, tightly coupled is used here to assist in differentiating clusters from grids. Armed with these definitions for clusters and grids, it is reasonable to view clusters as building blocks for grids. A cluster of clusters that are geographically dispersed can certainly be considered as grids.

6.5 CONCLUSION

The purpose of this chapter has been to frame clusters in the context of Grid Computing. The term cluster was defined, and a cluster landscape was introduced. Although this landscape revealed three distinct features, attention was focused here on HPC clusters. Based on the interaction of Smart System Software (SSS) with the underlying operating system, HPC clusters allow for two possibilities. At one end, Single System Image (SSI) SSS derives from a modification of the kernel of the operating systems. Targeted for the Linux operating environment, Beowulf BPROC provided the primary example for this case. In contrast, Single System Environment (SSE) SSS allows for HPC clustering via a layered-services approach that does not require kernel modification. This cross-platform solution allows for both capacity (throughput-driven) HPC and capability HPC clusters. Real-world examples illustrated that clustered architectures are delivering tangible and effective solutions today. Automating the design of integrated circuits and genome sequence analysis provided representative examples for capacity HPC. The embarrassingly parallel nature of the bioinformatics example also allowed introduction of desktop clustering via Platform ActiveCluster. Desktops are viewed as opportunistically available compute resources that pull workload for processing; these desktops compliment the dedicated workstations, servers, and supercomputers used in dedicated clusters based on Platform LSF. Parallel computing via MPI was illustrated through an automotive example. In this case, Platform HPC provided a representative example of capability HPC in an industrial manufacturing context. The process of federating clusters provides an excellent opportunity to investigate the similarities and differences between clusters and grids. Although clusters are definitely not grids, clusters do have grid-like affinities and are often the primary building blocks for grids. This comparison also permits a refinement of the cluster definition: "A cluster is a local-area, logical arrangement of independent entities that collectively provide a service."

In addition to HPC clusters, the cluster landscape reveals two other features:

- High-Availability (HA) clusters. HA clusters use smart system software (often referred to simply as middleware) to virtualize node-tied application instances to provide a highly available service. A classic example involves the provision of a highly available file service, e.g., NFS.
- Transactional clusters. Transactional clusters use software (e.g., round-robin IP name shuffling via DNS) or hardware (e.g., IP load balancing networking hardware) technology to virtualize instances at the IP packet level to provide a single point-of-presence. Traditional examples in this area include directory, name, mail, news, and Web Services.

Neither of these cluster types is currently given consideration in the context of Grid Computing. However, this situation is likely to change—particularly as OGSA and the utilitarian aspect of Grid Computing evolve. For example, HA has much in common with autonomic computing; experiences in HA clustering can be recontextualized for the grid. Although traditional transactional clusters will continue to provide services to grids (e.g., name resolution), emerging examples will more-directly incorporate transactional clusters. Platform Symphony, an integrated application and system infrastructure that supports the development, scheduling, and runtime execution of online/realtime distributed applications, provides insight into such possibilities. Taken together, clusters will continue to play a significant role in the ongoing evolution of Grid Computing.

7 | HPC Grids

Ian Lumb

Platform Computing, Inc.

In This Chapter

- HPC Architecture
- HPC Applications
- HPC Grids

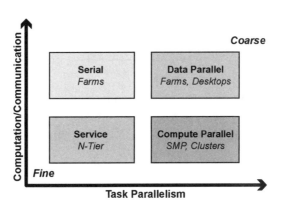

7.1 INTRODUCTION

Production High Performance Computing (HPC) was synonymous with scientific computing on "Big Iron" supercomputers. No longer dominated by just physical scientists and their *Grand Challenge Equations*, production HPC also embraces a variety of compute architectures.

BOX 7.1   What Are Grand Challenges?

What Are Grand Challenges?

Grand Challenge applications are fundamental problems in science and engineering with broad economic and scientific impact.   Grand Challenge projects generate increasingly complex data, require realistic simulations of the processes under study, and demand intricate visualizations of the results.   To meet these requirements, researchers need access to faster computers, more memory, disk space, and in many cases, multiple high-performance architectures across which their problems may be distributed.

They are generally considered intractible without the use of tremendous amounts of computing, memory, storage, and in some cases bandwidth resources.

Though framed in the broader context of non-traditional HPC, attention here focuses on parallel computing via the Message Passing Interface (MPI). Problems cast as MPI applications are seen to have a parallel-computing bias that reaches back into the numerical methods that have been used, and even to the originating science. Whereas MPI itself shows significant promise in addressing current-day computing challenges, in practice there are some serious shortcomings that need to be addressed for production HPC to be realized. Workload management system software allows the gap between MPI applications and their compute architectures to be closed, resulting in a solution for production HPC. A specific example of production HPC for the Linux operating environment shows that such solutions exist today. Moreover, the workload-management methodologies that apply at the cluster level have a natural affinity for extension to the grid. Overall, organizations are able to better empower the pursuit of science and engineering during MPI application development, deployment, and use.

7.2 FIVE STEPS TO SCIENTIFIC INSIGHT

To motivate the applications and architectures discussion, consider a scientific-inquiry example from the physical sciences.[1] Once the problem under investigation has been determined, the first task is to determine the relevant physics, chemistry, etc. (Figure 7.1, Step 1). This consideration results in a mathematical description of the problem that needs to be solved (Figure 7.1, Step 2). On the positive side, the mathematical description typically exists—i.e., there is rarely a need to invent the mathematical description. In many cases, the required mathematical description can be formulated by combining existing descriptions. Although the oldest and deepest explored of disciplines, and except in idealized situations subject to simplifying assumptions, mathematical solution methods are often not enough to allow the resulting equations to be solved. In mathematical terms, it is often difficult to near-impossible to derive *analytic* solutions to many scientific equations. To make matters worse, in some cases it is difficult to prove that such solutions even exist. Such existence theorems serve as a cornerstone for the practice of mathematics. Thus, science exposes serious mathematical challenges—in stark contrast to our childhood experiences with mathematics.

Given this challenging mathematical context, numerical methods are used to permit progress on otherwise unsolvable scientific problems (Figure 7.1, Step 3). Typically, this involves a discrete representation of the equation(s) in space and/or time, and performing calculations that trace out an evolution in space and/or time.[2, 3] It's important to note that the underlying structure of the resulting set of equations has impact on the types of numerical methods that can be applied.[4] Thus, numerical experiments are acts of modeling or simulation subject to a set of pre-specified constraints (Figure 7.1, Step 4). Problems in which time variations are key need to be seeded with *initial conditions*, whereas those with variations in space are subject to *boundary conditions*; it is not uncommon for problems to specify both kinds of constraints. The numerical model or simulation subject to various constraints can be regarded as a scientific application. Thus the solution of a scientific problem results in numerical output that may or may not be represented graphically (Figure 7.1, Step 5). One of four primary types (see section 7.3, "Applications and Architectures"), this application is further regarded as scientific workload that needs to be managed as the

---

[1] The analogous steps for informatics-heavy sciences such as the life sciences will be left for future consideration.

[2] Symbolic algebraic manipulation provides a notable analytic exception to this discrete approach.

[3] Maple, *http://www.maplesoft.com*

[4] There is an extensive literature base on this topic. Implementations of involving a collection of methods are often organized into libraries.

calculations are carried out. Because the practices of science and engineering are actually undertaken as a process of discovery, Figure 7.1 should be regarded as a simplifying overview that also fails to depict the recursive nature of investigation.

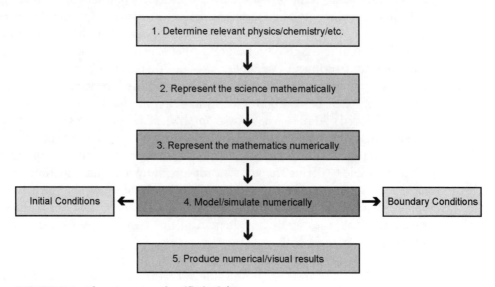

FIGURE 7.1 Five steps to scientific insight.

It may appear that numerical methods are the cure-all for any scientific problem that cannot be solved by mathematical methods alone. Unfortunately, that is not the case. On their own, many of the equations of classical physics and chemistry push even the most powerful compute architectures to their limits of capability. Irrespective of numerical methods and/or compute capability, these *Grand Challenge Equations* afford solutions based on simplifying assumptions, plus restrictions in space and/or time, etc. Because these particularly thorny equations are critical in science and engineering, there is an ongoing demand to strive for progress. Examples of Grand Challenge problems are provided elsewhere.[5]

7.3 APPLICATIONS AND ARCHITECTURES

Science dictates mathematics and mathematics dictates numerics[6] (Figure 7.1). Thus, there exists a numerics bias in all applications of scientific origin. This

---

[5] The Grand Challenge Equations, *http://www.sdsc.edu/Publications/GCequations*

[6] Numerics is used here as a shorthand for numerical methods.

predisposition motivates four types of applications (Figure 7.2) revealed by exploring process granularity. *Granularity* refers to the size of a computation that can be performed between communication or synchronization points.[7] Thus, any point on the vertical axis of Figure 7.2 identifies a specific ratio of computation (increasing from bottom to top) to communication (increasing from top to bottom).[8] Task parallelism, increasing from left-to-right on the horizontal axis, refers to the degree of parallelism present in the application. "Fine" through "Coarse" are used as qualitative metrics, as shown in the figure.

Most scientific problems are implemented initially as *serial* applications (Figure 7.2, Quadrant II[9]). These problems require that each step of the scientific calculation be performed in sequence. Serial applications can be executed on compute architectures ranging from isolated desktops, servers, or supercomputers to compute farms. Compute farms are loosely coupled compute architectures in which system software is used to virtualize compute servers[10] into a single system environment (SSE).

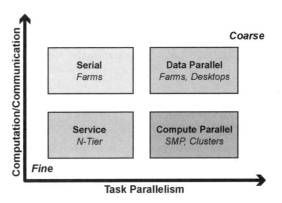

FIGURE 7.2 Applications and architectures.

---

[7] B. Wilkinson & M. Allen. *Parallel Programming: Techniques and Applications Using Networked Workstations and Parallel Computers*, Prentice Hall, Upper Saddle River, NJ, 1999.

[8] If the interest is computing, then communication is viewed as the "overhead" required to achieve the computation. Similarly, computation might be regarded as the overhead required to facilitate certain communications.

[9] The mathematical convention of numbering quadrants counter-clockwise from the upper-right-hand corner is used here.

[10] Although high-density, rack-mounted single/dual processor servers have been used in compute farms, there is an intensifying trend toward the use of higher-density blade servers in these configurations.

Various factors—e.g., time-to-results, overall efficiency, etc.—combine to demand performance improvements beyond what can be achieved by "legacy" serial applications alone. For those applications whose focus is on data processing, it is natural to seek and exploit any parallelism in the data. Such *data parallel* applications (Figure 7.2, Quadrant I) are termed *embarrassingly parallel* as the resulting applications are able to exploit the inherent coarse granularity. The parallelism in data is leveraged by Figure 7.3:

- Subdividing input data into multiple segments
- Processing each data segment independently via the same executable
- Reassembling the individual results to produce the output data

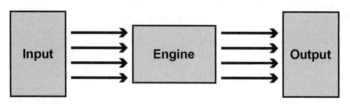

FIGURE 7.3   Data-driven parametric processing.

This data-driven approach accounts for one of four classes of parametric processing[11]. Although data-parallel applications are a solid match for isolated systems and compute farms, there is an increasing trend to harvest compute cycles from otherwise idle desktops. Regarded as opportunistic compute resources, desktops "pull" processing tasks from coordinating servers that they will execute as a background process—especially in place of a traditional screensaver. Desktops as compute elements gained initial popularity through the Peer-to-Peer (P2P) movement, and more recently in the context of Grid Computing. Ray-tracing applications provide a classic demonstration of this processing architecture. Despite the added challenge of managing data, successful implementations of embarrassingly parallel applications exist in many industries—e.g., genome sequencing in the life sciences, Monte Carlo simulations in High Energy Physics, reservoir modeling in petroleum exploration, risk analysis in financial services, etc. This approach is so appealing and powerful that it is often perceived to be of general utility. Unfortunately this simply isn't the case—and this is especially true for the Grand Challenge Equations identified previously (see section 7.2, "Five Steps to Scientific Insight").

---

[11] I. Lumb, B. Bryce, B. McMillan, and K. Priddy, *From the Desktop to the Grid: Parametric Processing in High-Performance Computing*, Sun User Performance Group, Amsterdam, Netherlands, October 2001.

If it exists at all, parallelism in the Grand Challenge Equations can be exploited at the source-code level—e.g., by taking advantage of loop constructs in which each calculation is independent of others in the same loop. Parallelism in data is absent or of minor consequence. This code-level parallelism lends itself to *compute parallel* (Figure 7.2, Quadrant IV) applications. Compute parallel applications are further segmented on the basis of memory access—i.e., shared versus distributed memory. With minimal language extensions and explicit code level directives, OpenMP[12] (and more recently Unified Parallel C (UPC)[13]), offer up parallel computing with shared-memory programming semantics. Symmetric MultiProcessor (SMP) systems allow for shared-memory programming semantics via threads[14] through uniform (UMA) and non-uniform (NUMA) memory access architectures.

Parallel Virtual Machine (PVM)[15] has given way to the Message Passing Interface (MPI)[16] as the standard for parallel computing with distributed-memory programming semantics. Likened to the "assembly language" for parallel programming[17] use of MPI requires a significant investment at the source-code level. In contrast to the use of threads in the shared-memory context, distributed processes are employed in the MPI case to achieve parallelism. MPI applications are typically implemented for tightly coupled compute clusters (see 7.4 "HPC Application Development Environment for additional details"). Although other factors (e.g., architecture access) need to be considered, numerics does influence the choice of parallel computing via shared versus distributed memory.[18] Both shared and distributed-memory parallel computing methodologies have been applied to scientific and engineering problems in a variety of industries—e.g., computational and combinatorial chemistry in the life sciences, computational fluid dynamics, crash simulations and structural analyses in industrial manufacturing, Grand Challenge problems in government organizations and educational institutions, plus sce-

---

[12] OpenMP, *http://www.openmp.org*

[13] Unified Parallel C, *http://upc.gwu.edu*

[14] In addition to the fork-and-exec creation of child processes, a parent process may also involve threads. Distinguishable by the operating system, threads can share or have their own memory allocations with respect to their parent process.

[15] Parallel Virtual Machine, *http://www.csm.ornl.gov/pvm/pvm_home.html*

[16] Message Passing Interface, *http://www.mpi-forum.org*

[17] K. Dowd & C. R. Severance, *High Performance Computing*, Second Edition, O'Reilly & Associates, Sebastopol, CA, 1998.

[18] Recent advances allow serial applications to be automatically enabled for MPI. By identifying code regions suitable for parallelization (e.g., repetitive calculations) via a templating mechanism, code-level modifications are applied. This approach is being applied extensively in the financial services area where numerical models change frequently.

nario modeling in financial services, etc. It is MPI compute parallel applications, traditional HPC, that will be the focus of attention here.[19]

Because MPI provides such a rich framework for computing in general, there are examples of MPI applications that communicate extensively while carrying out minimal processing (Figure 7.2, Quadrant III)—e.g., remote-to-direct-memory applications (RDMA), certain classes of search algorithms, etc. In addition, *service* applications whose focus is networking itself or Web Services[20] themselves would also fall into this area. As before, MPI applications would require tightly coupled architectures; whereas networking applications can be applied in a variety of contexts, loosely coupled architectures can be used in the instantiation of Web Services.

7.4 HPC APPLICATION DEVELOPMENT ENVIRONMENT

Together with the "commoditization" of low-processor-count, high-density servers, the emergence of low-latency, high-bandwidth interconnect technologies, MPI has played a key role in the widespread adoption of tightly coupled compute clusters for distributed memory-parallel computing:[21]

> *MPI is available everywhere and widely used, in environments ranging from small workstation networks to the very largest computers in the world, with thousands of processors. Every parallel computer vendor offers an MPI implementation, and multiple implementations are freely available as well, running on a wide variety of architectures. Applications large and small have been ported to MPI or written as MPI programs from the beginning, and MPI is taught in parallel programming courses worldwide.*

Applied through source-code-level modifications, MPI-specific directives are referenced against an MPI library at application link time. MPI libraries are architecture specific and may come from a variety of sources—e.g., a system vendor, an interconnect vendor or via an open source contribution. In each case, the relevant library implements the MPI specification[22] to some degree of compliance. This MPI library, in combination with the tools and utilities that support developers,

---

[19] Hybrid OpenMP-MPI applications allow scientists and engineers to simultaneously use threads and distributed processes when using a combination of SMP and clustered architectures.

[20] Web Services, *http://www.w3.org/2002/ws/*

[21] Gropp, W., E. Lusk, and A. Skjellum, *Using MPI: Portable Parallel Programming with the Message-Passing Interface*, Second Edition, The MIT Press, Cambridge, MA, 371 pp., 1999.

[22] Most MPI libraries fully implement version 1.x of the MPI specification, while many libraries are today supporting some subset of the version 2.x specification.

collectively forms the application development environment for a particular platform (Figure 7.4).

The situation described above might lead one to conclude that all of the requisites are present to smoothly enable MPI adoption. In practice, however, MPI has the following challenges:

- Re-synchronization and re-connection were not even factored in at the specification level.[23] There are no MPI implementations which account for this shortcoming. This is in striking contrast to PVM whose implementation allows for this.
- Fault tolerance was not factored in—even at the specification level;[23] again, there are no MPI implementations which account for this shortcoming. This can mean, for example, that an application can lose some of its processes, run to completion, and yield results of dubious validity.
- Hosts and numbers of processors need to be specified as static quantities—irrespective of actual usage conditions.
- Single point of control is absent. Although some recent MPI implementations offer a process daemon to launch MPI applications, there is little in the way of real application control.
- Multiple versions of MPI may exist on the same architecture. Applications need to carefully identify the relevant MPI libraries. The situation is more complex for MPI applications that span more than one execution architecture.

The upshot is clear: MPI places the responsibility of these shortcomings on the application developer and user. Because MPI applications are challenging to control and audit, a better production HPC solution is required.

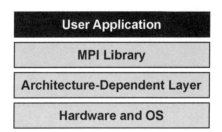

FIGURE 7.4 MPI application development environment.

---

[23] K. Dowd and C. R. Severance, *High Performance Computing*, Second Edition, O'Reilly & Associates, Sebastopol, CA, 446 pp., 1998.

7.5 PRODUCTION HPC REINVENTED

There exists a gap between the potential for distributed-memory parallel computing via MPI and what is actually achievable in practice. The use of system software
allows this gap to be closed, and the promise of MPI to be fully realized. In the
process, the notion of production HPC is redefined. To start, consider a modified
version of Figure 7.4 in which the newly added workload management system software is shown in Figure 7.5.

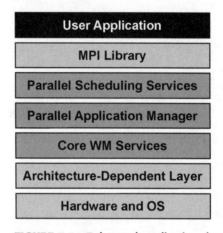

FIGURE 7.5 Enhanced application development
environment via workload management system software.

Figure 7.5 introduces the following three components to the MPI application
development environment:

- Core workload management services—This system software component allows a heterogeneous collection of compute servers, each running their own instance of an operating system, to be virtualized into a compute cluster.[24, 25]
  Sensory agents are used to maintain static and dynamic information in real
  time across the entire compute cluster. Primitives for process creation and
  process control across a network are also provided.
- Parallel application management—Challenges specific to the management of
  MPI parallel applications includes the need to:

---

[24] This layered-services approach has been contrasted with Beowulf clustering (via a distributed
process space) elsewhere.

[25] Lumb, I., "Linux Clustering for High-Performance Computing," *;login:*, USENIX/SAGE, **26**(5), 79-
84, August 2001.

- Maintain the communication connection map.
- Monitor and forward control signals.
- Receive requests to add, delete, start, and connect tasks.
- Monitor resource usage while the user application is running.
- Enforce task-level resource limits.
- Collect resource usage information and exit status upon termination.
- Handle standard I/O.
- Parallel scheduling services—Workload management solutions typically employ a policy center to manage all resources—e.g., jobs, hosts, interconnects, users, queues, etc. Through the use of a scheduler, and subject to predefined policies, demands for resources are mapped against the supply of resources in order to facilitate specific activities. Scheduling policies of particular relevance in parallel computing include advance reservation, backfill, preemption, processor, and/or memory reservation, etc.

The combined effect of these three layers of a workload management infrastructure allow the shortcomings of MPI to be directly addressed:

- Absence of re-synchronization and re-connection—Although the workload management infrastructure (Figure 7.5) cannot enable re-synchronization or re-connection, by introducing control across all of the processes involved in an MPI application, there is greatly improved visibility into synchronization and/or connection issues.
- Absence of fault tolerance—At the very least, the introduction of a workload management infrastructure provides visibility into exceptional situations by trapping and propagating signals that may be issued while workload is executing. These signals can be acted upon to automatically re-queue workload that has thrown an undesirable exception. Even better, when integrated with the checkpoint/restart infrastructure of a workload manager, interrupted workload can continue to execute from the last successful checkpoint—often without user intervention.
- Absence of load balancing—The need to explicitly identify hosts and numbers of processors can be regarded as an absence of load balancing—i.e., the specification of resources that ignores current resource-usage conditions. Because workload-management infrastructures maintain dynamic load state across the entire compute infrastructure in real time, this shortcoming is completely eliminated.
- Absence of single point of control—A parallel application manager provides a single point of control. This ensures that all the distributed processes that collectively comprise the MPI application are managed and accounted for. Again a key shortcoming becomes a core competence via workload-management system software.

- Multiple versions of MPI—Through configuration information, MPI applications are able to specify the relevant MPI library. A very flexible architecture also allows the workload-management system software to work in concert with the existing parallel application development and runtime environment.

Together with the existing parallel application development and run-time environment, workload-management system software allows the inherent shortcomings of MPI to be effectively eliminated. The cumulative effect is to practically reinvent HPC via MPI. This threefold reinvention is captured in Figure 7.6. Starting in the upper-right of this figure, MPI applications are developed in-house or acquired from commercial providers. On job submission, these applications are accompanied by a description of their run-time resource requirements—e.g., a processor-count range,[26] data-management directives (e.g., a file transfer), plus other environmental requirements.[27] The scheduling agent takes into account the pre-specified resource requirements, the unique characteristics of the compute architectures that it has available, and subject to the policies that reflect organizational objectives. Because the scheduler creates a run-time environment for the workload, its task is one of dynamic provisioning. On dispatch, the scheduler has optimally matched the workload to an appropriate run-time architecture subject to established policies. The application executes in a fully managed environment. A

---

[26] Experience dictates that every parallel application shows acceptable performance characteristics over a range of processors. A typical criterion is that the performance remain close to linear as the number of processors increases. This is referred to as linear speedup. Effective workload management systems allow this processor count to be specified as a range at workload submission time. This serves to automate the load-balancing situation, and to enhance to overall effectiveness of the scheduling services.

[27] The need to bind processes to processors serves as one example of an execution environment requirement. In such cases, the workload-management infrastructure works in tandem with the operating system, interconnect manager, etc., to address the requirement.

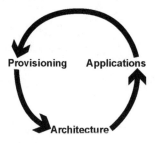

FIGURE 7.6 Production HPC reinvented.

comprehensive audit trail ensures that all activities are accounted for. In this way, there exists a closed loop for production HPC—one that enhances the developer and user experience rather than encumbering it. A specific solution example is provided in the following section.

7.6 HPC GRIDS

It has been suggested that production HPC can be reinvented through the use of an integrated development and run-time environment in which workload management system software plays a key role. A complete example is considered here to further illustrate this reinvention. Consider the integrated production HPC solution stack shown in Figure 7.7. Although this example is based on the Linux operating environment, similar stacks for other operating environments can be crafted.

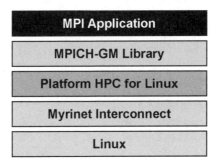

FIGURE 7.7 Production HPC for Linux.

At the base of the production HPC solution stack for Linux are low-processor-count servers[28] each running their own instance of the GNU/Linux operating system. Ethernet-based network interface cards (NICs) allow for standard TCP/IP based services between these systems. Because TCP/IP over Ethernet necessitates assumptions regarding shared access and the occurrence of collisions, the result is a communications protocol that is latency heavy and therefore inefficient for message-passing in support of parallel computing. Hence each system implements the

---

[28] Historically, alpha-based processors were used due to their excellent floating-point-performance characteristics. With the advent of first-generation Itanium Processor Family CPUs, it is expected that 64-bit Intel Architecture (IA-64) will eventually dominate in this space. In the interim, fourth-generation IA-32 Pentium Processor Family CPUs hold the price/performance niche.

GM message passing protocol across low-latency, high-bandwidth, multi-port Myricom Myrinet switches[29] used solely to support parallel computing via MPI and the GM driver. By identifying each available Myrinet port as a resource to the core workload management services provided by Platform HPC for Linux,[30] the provisioning component of this infrastructure is aware of the static and dynamic attributes of the architecture it can apply parallel scheduling policies against. Platform HPC for Linux also provides the control and audit primitives that allow parallel applications to be completely managed. User's MPI applications need to be compiled and linked against the application development environment provided by Myricom. This ensures that the appropriate GM-protocol modifications to the widely adopted open source MPICH implementation[31] of MPI are used. Through configuration information, the workload management system software based on Platform HPC for Linux is made aware of the enhanced MPI run-time environment. (For more information on MPICH, see Chapter 6.)

Portability was identified as a design goal for MPI. This objective has been carried through in MPI implementations such as MPICH. Despite this fact, heterogeneous parallel applications based on MPI must not only use the same implementation of MPI (e.g., MPICH) but also the same protocol implementation (e.g., GM). In other words, MPI is a multi-protocol API (Figure 7.7) in which each protocol implements its own message formats, exchange sequences, etc.

Because they can be regarded as a cluster of clusters, it follows that computational grids might provide a suitable run-time environment for MPI applications. The fact that grids are concerned with resource aggregation across geographic (and other) domains further enhances the appeal. Fortunately, Platform HPC for Linux is consistent with Platform MultiCluster—system software that allows independent clusters each based on Platform LSF to be virtualized into an Enterprise Grid.

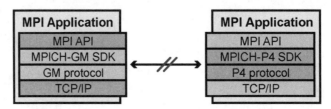

FIGURE 7.8 The multi-protocol nature of MPI.

---

[29] Myricom Myrinet, *http://www.myricom.com/myrinet*

[30] Platform Computing Corporation, "Using LSF with MPICH-GM," Technical Documentation, April 2002, 10 pp.

[31] MPICH-G2, *http://www.hpclab.niu.edu/mpi*

This combination introduces the possibility for exciting new modes of infrastructural provisioning through various grid-centric scheduling policies—e.g., Grid Advance Reservation, Grid Fairshare, and Grid Resource Leasing.

Of these grid-centric policies, Grid Resource Leasing (GRL) is particularly novel and powerful. GRL allows sites to make available fractions of their resources for use by other sites participating in their Enterprise Grid. These collective resources can be occupied by MPI applications through requirements specifications. In this fashion, users can *co-schedule* MPI applications to span more than one geographic location. Even more impressive is the fact that this co-scheduling is automatically enabled at run time—i.e., existing MPI applications do not need to be relinked with special libraries. By jointly leveraging the parallel application management capability already present in Platform HPC for Linux, in concert with this grid-level scheduling policy of Platform MultiCluster, MPI application users take advantage of their entire Enterprise Grid in a transparent and effective fashion. Platform MultiCluster and its grid-level scheduling policies are considered in detail elsewhere.[32]

7.7 CONCLUSION

Production High Performance Computing (HPC) incorporates a variety of applications and compute architectures. Widespread use of the Message Passing Interface (MPI) is better enabled through the use of workload management system software. This software allows MPI applications and the compute architectures on which they execute to be provisioned on demand. This critical link significantly reduces complexity for the scientist or engineer, thus reducing time to results, and ensuring overall organization efficiency. A tightly coupled cluster based on the Linux operating environment was shown to be a particularly attractive and viable compute architecture. The incorporation of this environment into a compute grid was also shown to be a natural progression. Overall, organizations are able to better empower the pursuit of science and engineering during application development, deployment, and use.

Acknowledgments

Chris Smith, Brian MacDonald, Bill Bryce, and Bill McMillan are all acknowledged for their contributions to this article.

---

[32] Platform Computing Corporation, "Platform MultiCluster Guide," Technical Documentation, June 2002, 82 pp.

8 Data Grids

Andrew Grimshaw
Avaki Corporation

In This Chapter

- Data Grids Defined
- Data Grid Architecture

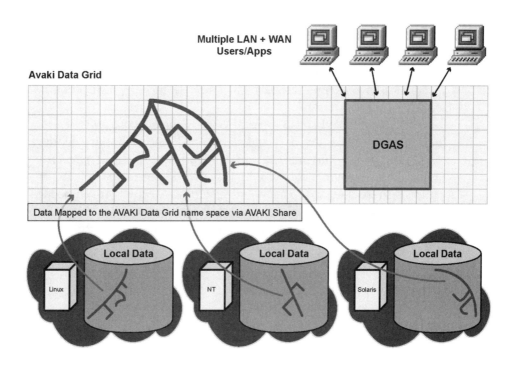

8.1 INTRODUCTION

> *For over thirty years science fiction writers have spun yarns featuring world-wide networks of interconnected computers that behave as a single entity. Until recently such science fiction fantasies have been just that. Technological changes are now occurring which may expand computational power in the same way that the invention of desk top calculators and personal computers did. In the near future computationally demanding applications will no longer be executed primarily on supercomputers and single workstations using local data sources. Instead enterprise-wide systems, and someday nationwide systems, will be used that consist of workstations, vector supercomputers, and parallel supercomputers connected by local and wide area networks. Users will be presented the illusion of a single, very powerful computer, rather than a collection of disparate machines. The system will schedule application components on processors, manage data transfer, and provide communication and synchronization in such a manner as to dramatically improve application performance. Further, boundaries between computers will be invisible, as will the location of data and the failure of processors.*[1]

The future is now; after almost a decade of research and development by the grid community, we see grids (then called metasystems[2]) being deployed around the world both in academic settings, and more tellingly, in production commercial settings.

What is a grid? What use is a grid? What is required of a grid? Before we answer these questions, let us step back and define "grid" and look at its essential attributes.

Our definition, and indeed a popular definition, is: A grid *system* is a collection of distributed resources connected by a network. A grid system, also called a *grid*, gathers resources—desktop and hand-held hosts, devices with embedded processing resources such as digital cameras and phones, or tera-scale supercomputers—and makes them accessible to users and applications in order to reduce overhead and accelerate projects. A grid *application* can be defined as an application that operates in a grid *environment* or is "on" a grid system. Grid system software (or *middleware*) is software that facilitates writing grid applications and manages the underlying grid infrastructure.

The resources in a grid typically share at least some of the following characteristics:

---

[1] A.S. Grimshaw, "Enterprise Wide Computing," *Science*, 256: 892–894, Aug 12, 1994.

[2] L. Smarr and C.E. Catlett, "Metacomputing," *Communications of the ACM.* 35(6):44–52, June 1992.

- They are numerous.
- They are owned and managed by different, potentially mutually distrustful organizations and individuals.
- They are potentially faulty.
- They have different security requirements and policies.
- They are heterogeneous, i.e., they have different CPU architectures, are running different operating systems, and have different amounts of memory and disk.
- They are connected by heterogeneous, multilevel networks.
- They have different resource management policies.
- They are likely to be separated geographically (on a campus, in an enterprise, on a continent).

A grid enables users to collaborate securely by sharing *processing*, *applications*, and *data* across systems with the above characteristics in order to facilitate collaboration, faster application execution, and easier access to data. More concretely, this means being able to:

- Find and share data—When users need access to data on other systems or networks, they should simply be able to access them like data on their own system. System boundaries that are not useful should be invisible to users who have been granted legitimate access to the information.
- Find and share applications—The leading edge of development, engineering, and research efforts consists of custom applications—permanent or experimental, new or legacy, public-domain or proprietary. Each application has its own requirements. Why should application users have to jump through hoops to get applications together with the data sets needed for analysis?
- Share computing resources—It sounds very simple—one group has computing cycles, some colleagues in another group need them. The first group should be able to grant access to its own computing power without compromising the rest of the network.

Grid Computing is in many ways a novel way to construct applications. It has received a significant amount of recent press attention and been heralded as the next wave in computing. However, under the guises of "peer-to-peer systems," "metasystems," and "distributed systems," Grid Computing requirements and the tools to meet these requirements have been under development for decades. Grid Computing requirements address the issues that frequently confront a developer trying to construct applications for a grid. The novelty in grids is that these requirements are addressed by the grid infrastructure in order to reduce the burden on the application developer. The requirements (described more thoroughly else-where[3, 4]) are:

- security
- a global name space
- fault-tolerance, accommodating heterogeneity
- binary management
- multi-language support
- scalability
- persistence
- extensibility
- site autonomy
- complexity management

Solving these requirements is the task of a grid infrastructure. An architecture for a grid based on sound design principles is required in order to address each of these requirements. In this chapter, we will focus on one particular aspect of grids—data.

8.2 DATA GRIDS

Data grids are used to provide secure access to remote data resources: flat-file data, relational data, and streaming data. For example, two collaborators at sites A and B need to share the results of a computation performed at site A, or perhaps design data for a new part needs to be accessible by multiple team members working on a new product at different sites—and in different companies.

We will examine in detail a grid created using Avaki Data Grid software (herein referred to as a data grid). A data grid provides transparent, secure, high-performance access to federated data sets across administrative domains and organizations. Users (both people and applications) may be unaware that they are using a data grid. We begin with an examination of alternatives to data grid solutions. We then look at Avaki Data Grid (ADG) in detail, starting with a close examination of the design principles and then the overall architecture.

---

[3] A.S. Grimshaw and W.A. Wulf, "The Legion Vision of a Worldwide Virtual Computer," *Communications of the ACM*, 40(1): 39–4, Jan 1997.

[4] A.S. Grimshaw, A. Natrajan, M.A. Humphrey, M.J. Lewis, A. Nguyen-Tuong, J.F. Karpovich, M.M. Morgan, A.J. Ferrari, "From Legion to Avaki: The Persistence of Vision," *Grid Computing: Making the Global Infrastructure a Reality*, eds. Fran Berman, Geoffrey Fox and Tony Hey, 2003.

8.3 ALTERNATIVES TO DATA GRIDS

The problems that data grids solve have been around for as long as networks have existed between computers. A number of solutions have been created to handle the problems of remote data access. In this section, we take a closer look at some of the more popular solutions and present their advantages and disadvantages. For each solution, we will take up a case of a local user at one site trying to share a file with a remote user at another site. The first user, Alice, is a user on the local machines owned by one company, whereas the second user, Bob, is a user on the remote machines owned by another company. The two companies may not share a mutually trustful relationship although the sharing of Alice's file with Bob has been approved.

8.3.1 Network File System (NFS)

NFS is the standard Unix solution for accessing files on remote machines within a LAN. With NFS, a disk on a remote machine can be made part of the local machine's file system. Accessing data from the remote system now becomes a matter of accessing a particular part of the file system in the usual manner. In our use case above, Alice could run an NFS server on her machine and Bob could run an NFS client to mount Alice's file system onto his. Bob can now access the exact file that Alice wishes to share.

There are several advantages to NFS, one of the most significant of which is that it is easy to understand. Typically, Unix system administrators configure the server and client, and ordinary users such as Alice and Bob simply use it without necessarily realizing that they are doing so. Moreover, applications need not be changed to access files on an NFS mount—the NFS server supports standard OS file system calls. Accordingly, files may be accessed entirely or in parts, as desired. Finally, the NFS server and client tools come standard on all Unix systems. On Windows, a special service pack must be purchased and installed.

The biggest disadvantage with NFS is that it is a LAN protocol—it simply does not scale to WAN environments. If Alice and Bob are separated by more than a few buildings, using NFS between them is not viable. Moreover, if Alice and Bob belong to different organizations, as they do in our use case, NFS cannot be deployed with reasonable guarantees of security.

Three characteristics of NFS doom it for use in wide-area, multi-organizational settings. First, the caching strategy on the NFS server typically releases data after 30 seconds and reloads the data on subsequent access. The result is a frequent retransmission of data and over-consumption of bandwidth. A related problem is that the read block size is too small, typically 8KB. In a wide-area environment, latency can be high, therefore larger block sizes are needed to amortize the cost of the remote procedure call (RPC). Although the block size can be changed, most NFS clients do not change it.

Second, and most seriously, NFS does not address security well. An NFS request packet is sent in the clear and contains the (integer) User ID (UID) and Group ID (GID) of the user making the read or write request. The NFS server "trusts" the NFS client to not lie about the identity of the user making the request. Such a trustful relationship does not exist among multiple organizations, such as Alice's and Bob's. Even if the organizations do trust each other, man-in-the-middle, imposter, and snooping attacks can be made with NFS traffic. A Virtual Private Network (VPN) deployed between the organizations may attenuate some of these attacks, but VPNs introduce their own problems of management, trust, and scalability. Further, firewalls typically do not permit NFS traffic through them, making NFS impossible for cross-site use across firewalls.

Third, even assuming that the packets can be sent in a safe and trustworthy fashion, NFS requires that the identity spaces at the two sites to be the same. In other words, not only should Alice and Bob have accounts on each other's machines, but Alice's UID on Bob's machine must be the same as her UID on her own machine. Likewise, Bob must synchronize his UIDs on his machine and Alice's machine in the same manner. Such synchronization would be possible if Alice and Bob were within a single domain; in our realistic use case, they are not.

Other disadvantages plague NFS; we will mention them briefly here. NFS performance does not scale in a wide-area setting because it is a request-reply protocol, which requires acknowledgments to be sent for every request, thus increasing effective transmission latency. NFS is a stateless protocol, i.e., the server does not keep track of the position of files being read. Accordingly, the server cannot pre-cache data or pre-position accesses to give clients better performance. Increasing the number of clients overwhelms the one server deployed to serve data, thus reducing performance. In our use case, Alice wants to give Bob access to one of her files. If Bob also had some files he wished to share with Alice, he would have to run an NFS server on his machine and ask Alice to run an NFS client on hers. This kind of configuration can lead to a morass of cross-mounting, which can overburden most administrators. In general, NFS requires $m \times n$ connections if m clients access data on n servers.

8.3.2 File Transfer Protocol (FTP)

FTP has been the tool of choice for transferring files between computers since the 1970s.[5] FTP is a command-line tool that provides its own command prompt and has its own set of commands. Several of the commands resemble Unix commands, although several new commands, particularly for file transfer as well manipulating

---

[5] FTP specification, *http://www.ietf.org/rfc/rfc0765.txt?number=765*

the local file system, are different. FTP may be used within a script; however, in that case, the password for the remote machine must be stored in a clear-text file on the local machine. Using ftp, Alice may connect to Bob's machine, enter a username and password relevant to Bob's machine, change to the appropriate remote directory and then transfer the file.

The benefits of using ftp is that it is relatively easy to use, has been around for a long time, and is therefore likely to be installed virtually everywhere. However, the disadvantages of ftp are numerous. First, Alice must have access to an account on Bob's machine, complete with username and password. Having such access means that Alice potentially could do more than just file transfer—she might be able to log in to Bob's machine and access files, directories, and other machines to which she has not been given explicit access. From Alice's perspective, every transfer requires her typing the appropriate machine name, username, and password. She could ameliorate some of this burden by using a configuration file for ftp, but that file may require storing a clear-text password for Bob's machine.

In order to eliminate some of these problems, Bob's site may choose to implement anonymous ftp. In this case, Alice need not have a username and password for Bob's machine, but must still remember the machine name and part of the directory structure. The problem with anonymous ftp is obvious—*anyone* may now access Bob's ftp directory, not just Alice. The potential for unauthorized overwriting or filling up of disk space is large.

FTP is also inherently insecure; passwords and data are transmitted in the clear. Snooping attacks may easily compromise Alice and Bob. Hence, most sites that have firewall protection shut down the standard ftp port to discourage such attacks, making ftp unfeasible. Even without firewalls, there are other disadvantages to using ftp. Because ftp requires making a copy of the data on Bob's machine, if Alice ever changes her own copy of the file, she must remember to ftp the new version of the file. Moreover, if Bob ever changes the file, he must remember to ftp the file back to Alice and reconcile concurrent changes, if any. This process is fraught with the potential for inconsistencies and the problem is compounded if additional people need to use Alice's file and receive versions from her at different points in time. Also, ftp is an all-or-nothing protocol; if even one bit of a large file changes, the entire file must be copied over. Finally, ftp is not conducive to programmatic access. Applications cannot use ftp to take advantage of remote files without significant modification.

8.3.3 NFS over IPSec

IPSec is a protocol devised by IETF to encrypt data on a network. With IPSec installed and configured properly, all traffic on a network can be encrypted. Consequently, illegitimate snooping of network traffic does not affect the privacy and integrity of the communication between a server and a client. NFS over IPSec implies traffic between an NFS server and an NFS client over a network on which the data has been encrypted using IPSec. The encryption is transparent to an end-user. NFS over IPSec removes some, but not all, of the disadvantages of using NFS.

NFS over IPSec results in encrypted NFS traffic, thus regaining privacy and integrity. However, NFS continues to be a LAN-based protocol that does not scale to the WAN-like environment typical in our use case. All of the performance, scalability, configuration, and identity space problems we discussed earlier remain. In addition, in order to deploy IPSec, all of the machines in Alice's and Bob's domains must be reconfigured. Specifically, their kernels must be recompiled in order to insert IPSec in the communication protocol stack. This recompilation is hard, and anecdotal evidence suggests that the recompilation is risky, error-prone, and ill-documented. Finally, once this recompilation is done, *all* traffic between all machines is encrypted. Even Web, e-mail, and ftp traffic is encrypted, whether desired or not.

8.3.4 Secure Copy—scp/sftp

SCP/SFTP belong to the *ssh* family of tools. SCP is basically a secure version of the Unix rcp command that can copy files to and from remote sites, whereas sftp is a secure version of ftp. Both are command-line tools. The syntax for scp resembles the standard Unix cp command with a provision for naming a remote machine and a user on it. Likewise, the syntax and usage for sftp resembles ftp.

The benefits of using scp/sftp are that their usage is similar to existing tools. Moreover, password and data transfer is encrypted, and therefore secure. However, a disadvantage is that these tools must be installed specifically on the machines on which they will be used. Installations for Windows are hard to come by. Moreover, scp/sftp do not solve several of the problems with ftp. In our use case, Alice must still have access to an account on Bob's machine. Alice must continue to remember the appropriate machine name, username, and password. She could ameliorate some of this burden by using an authorized keys file which permits password-less access, but she must then store her private key safely on her local machine.

Sites protected by firewalls may permit scp/sftp traffic on the designated port because the traffic is encrypted. However, scp/sftp does not attempt to solve the consistency problems of proliferating multiple copies of the file. Like ftp or rcp, a change of even one bit requires the entire file to be copied over. Finally, these tools are not conducive to programmatic access. Applications cannot take advantage of remote files using scp/sftp without significant modification.

8.3.5 De-Militarized Zone (DMZ)

A *DMZ* is simply a third set of machines accessible to both Alice and Bob using ftp or scp/sftp, established to create an environment trusted by both parties. When Alice wishes to share a file with Bob, she must transfer the file to a machine in the DMZ, inform Bob about the transfer, and request Bob to transfer the file from the DMZ machine to his own machine. Although both Alice and Bob have relatively unfettered access to the DMZ machines, neither party compromises his/her own machine by letting the other have access to it.

With a DMZ, neither Alice nor Bob requires an account on the other's machine. Typically, companies deploying DMZs also deploy scp/sftp or some such secure means of file transfer. Therefore, these tools must be installed on all concerned machines. Alice and Bob both have to remember machine names, usernames, and passwords for the DMZ machines. However, now they also have to remember an additional step of informing the other whenever a transfer occurs.

DMZs worsen the consistency problems by maintaining three copies of the file. Also, because the file essentially makes two hops to get to its final destination, network usage increases. DMZs may address security concerns, but they do not ameliorate any of the other problems with scp/sftp and they do increase administrative burden. If Alice's company decides to cooperate with a third company, thus requiring Alice to interact with Chris at that company, she must now create and remember yet another DMZ configuration for interacting with Chris. The Alice-Bob DMZ cannot be reused because of the potential for Chris to access files intended for Bob.

8.3.6 GridFTP

GridFTP is a tool for transferring files. It is built on top of the Globus Toolkit.[6] GridFTP is an example of a service that characterizes the Globus "sum of services" approach for a grid architecture. Alice and Bob, in our use case, could use GridFTP to transfer files from one machine to another, similar to the way they would use ftp. Naturally, both parties must install the Globus Toolkit in order to use this service.

GridFTP solves the privacy and integrity of the problems with ftp by encrypting passwords and data. Moreover, GridFTP provides for high-performance, concurrent accesses by design. An API enables accessing files programmatically, although applications must be rewritten to use new calls. Data can be accessed in a variety of ways—for example, blocked and striped. Part or all of a data file may be accessed, thus removing the all-or-nothing disadvantage of ftp.

---

[6] W. Allcock, J. Bester, J. Bresnahan, A. Chervenak, L. Liming, S. Meder, S. Tuecke, "GridFTP Protocol Specification," *GGF GridFTP Working Group Document*, September 2002.

However, GridFTP does not address the identity space problems with ftp. Alice and Bob in our use case must still have an account on each other's machine, thus giving them more privileges than just file access. Instead of a machine name, username, and password as in ftp, Alice and Bob have to remember just the machine name. Their identities are managed by Globus, using session-based credentials. Finally, GridFTP does not solve the problems of maintaining consistency between multiple copies, because Alice and Bob would still be required to maintain at least two copies of the file, one on each user's machine.

8.3.7 Andrew File System (AFS)

The *Andrew File System* is a distributed network file system that enables access to files and directories distributed across multiple sites. Access to files involves becoming part of a single virtual file system. AFS comprises several cells, with each cell representing an independently administered file system. In our use case, the file system on Alice's machine would be one cell, whereas the file system on Bob's machine would be another. The cells together form a single large virtual file system that can be accessed similar to a Unix file system.

AFS permits different cells to be managed by different organizations, thus managing trust. In our use case, Alice and Bob would not require accounts on the other's machines. Also, they could control each other's access to their cell using the fine-grained permissions provided by AFS. When Bob accesses one of Alice's files for which he has permission, he accesses exactly the current copy of the file. Thus, AFS avoids the consistency problems with other approaches using copy-on-open semantics unless there are multiple concurrent writers (which AFS does not deal with well.) In order to improve performance, AFS supports intelligent caching mechanisms. Because access to an AFS file system is almost identical to accessing a Unix file system, users have to learn few new commands, and legacy applications can run almost unchanged.

AFS implements strong security features. All data are encrypted in transit. Authentication is using Kerberos, and access control lists are supported.

The drawbacks of AFS revolve around the use of Kerberos and the fact that it is a file system. By way of explanation, the use of Kerberos means that *all* sites and organizations that want to connect using AFS must themselves use Kerberos authentication *and* all of the Kerberos realms must trust each other. In practice, this means changing the authentication mechanism in use at the organization. This is a nontrivial—and typically politically very difficult—step to accomplish. Second, the realms must trust each other. This is similarly difficult to accomplish. Third, the Kerberos security credentials time-out eventually. Therefore, long-running applications must be changed to renew credentials using Kerberos's API. Also, AFS requires that all parties migrate to the same file system. In other words, Alice and Bob would have to migrate their entire file systems to AFS, which would probably be a significant burden on them and their organizations.

8.4 AVAKI DATA GRID

The objective of Avaki Data Grid is to provide high-performance; easy, transparent, secure collaboration; and coherent sharing between different locations, administrative domains, and organizations. Let's look briefly at each of these objectives.

- High-performance—Nobody wants a low-performance system. Yet remote access is inherently slower than local access due to the combination of higher latency and often lower bandwidth. To provide high-performance access in the wide-area, local copies must be cached to reduce the time spent transferring data over the wide-area network.
- Coherent—Caching data is great for performance. Unfortunately, it can lead to inconsistent copies of the data, which can lead in turn to incorrect application results or bad decisions based on out-of-date data. Thus, the data grid must provide cache-coherent data while recognizing *and* exploiting the fact that different applications have different coherence requirements.
- Transparent—The data grid must be transparent to end users and applications. If users have to change their code or behaviors in order to use the data grid, then they are less likely to use it—reducing the benefit of having the grid in place.
- Secure—"Secure" is a word that covers a wide range of issues. Many users believe that a data grid must support strong authentication with identities that span administrative domains and organizations, support the establishment of virtual organizations (groups that span organizations), enforce access control policies, and protect data.
- Between different administrative domains and organizations—To span administrative domains, a grid must address the identity mapping problem. To span organizations issues of trust management must be addressed.

In designing Avaki Data Grid to meet these goals, three design principles were kept in mind:

- Provide a single-system view—With today's operating systems, we can maintain the illusion that our local area network is a single computing resource. But once we move beyond the local network or cluster to a geographically dispersed group of sites, perhaps consisting of several different types of platforms, the illusion breaks down. Researchers, engineers, and product development specialists (most of whom do not want to be experts in computer technology) must request access through the appropriate gatekeepers, manage multiple passwords, remember multiple protocols for interaction, keep track of where everything is located, and be aware of specific platform-dependent limitations (e.g., this file is too big to copy, or to transfer to one's system; that application runs only on a certain type of computer). Re-creating the illusion of single resource for heterogeneous, distributed resources reduces the complexity of the overall system and provides a single namespace.

- Provide transparency as a means of hiding detail—Grid systems should support the traditional distributed system transparencies: access, location, heterogeneity, failure, migration, replication, scaling, concurrency, and behavior. For example, a user or programmer should not have to know where a peer object is located in order to use it (access, location, and migration transparency), nor does a user want to know that a component across the country failed—they want the system to recover automatically and complete the desired task (failure transparency). This is the traditional way to mask various aspects of the underlying system.

- Reduce "activation energy"—One of the typical problems in technology adoption is getting users to use it. If shifting to a new technology is difficult, then users will tend not to make an effort to try it unless their need is immediate and extremely compelling. This is not a problem unique to grids—it is human nature. Therefore, one of our most important goals is to make using the technology easy. Using an analogy from chemistry, we keep the activation energy of adoption as low as possible. Thus, users can easily and readily realize the benefit of using grids—and get the reaction going—creating a self-sustaining spread of grid usage throughout the organization. Another variant of this concept is the motto "no play, no pay." The basic idea is that if you do not need a feature, e.g., encrypted data streams, fault resilient files, or strong access control, you should not have to pay the overhead of using it.

Avaki Data Grid meets our goals using a federated sharing model, a global name space, and a set of servers—called DGAS (Data Grid Access Servers) that support the NFS protocols and can be mounted by user machines, effectively mapping the data grid into the local file system. Let's break this down.

Start with the global name space. This is really a fancy word for a globally visible directory structure where the leaves may be files, directories, servers, users, groups, or any other named entity in the data grid. Thus, the path "/shares/grimshaw/my-file" uniquely identifies `myfile`, and the path can be used anywhere in the data grid by a client (user or application) to refer to `myfile` regardless of where the client is located, and regardless of where `myfile` is located.

The data get into the data grid—and get a path name in the global name space—when they are *shared*. The share command takes a rooted directory tree on some source machine and maps it into the global name space. For example, the user can share c:\data on his laptop into /shares/grimshaw/data. At that point, the data on the laptop \data directory are available for both read and write, subject to access control, by any authorized user in the grid.

Data access in ADG is, for the most part, via the local file system on the user's machine, which in turn is an NFS client to a DGAS. The DGAS looks to the local operating system file system like a standard NFS 3.0 server, though that is not, in fact, the case. End users are unaware that they are even using the ADG; their shell scripts, Perl scripts, and other applications that use *stdio* will work without any modification on the ADG. We choose NFS because most operating systems have a native NFS client.

Access control in ADG is via access control lists (ACLs) on each grid object (file, directory, share, group, etc.). For each operation on an object (e.g., read, write) there is an *allow* list and a *deny* list. The allow list is first evaluated. If a request is not from a user in the allow list, the operation is rejected. Then, if the user is allowed, the deny list is checked. The deny list overrides the allow list. The lists may contain either individuals or groups. Groups are sets of individuals—perhaps in different organizations. The use of cross-organization groups facilitates the definition of virtual organizations.

In the example in Figure 8.1 we have shared three different directory structures into the grid: one each from a Solaris machine, a Windows NT machine, and a Linux machine. ADG does not care whether the data is stored on direct attached disk as in my example above, on Network Attached Storage (NAS), on a Storage Area Network (SAN), or perhaps even on optical media. Furthermore, the data stays where it is. That means that local applications that count on the data in a particular directory can still access the data, and all local backup and other data management procedures continue to function.

Modifications to shared data by either direct means on the host machine, or via ADG, are visible to both, though in the case of subsequent ADG access the coherence window applies (more on this in section 8.5).

Given the above example, let's examine how the data grid is viewed both by the end user and by the system administrator.

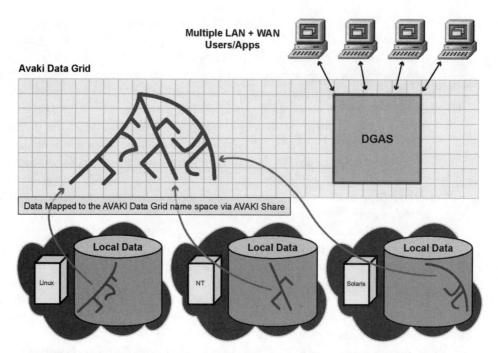

FIGURE 8.1 Data at three different sites, on three different types of machine have been mapped into the data grid. Data can now be accessed from anywhere in the grid. Typically access is via a Data Grid Access Server (DGAS). The DGAS appears to local operating systems as an NFS 3.0 server, providing standardized, secure, and transparent access to data.

8.4.1 Accessing the Data Grid

The first thing to stress about the user's view of the ADG is that no programming is required at all. Applications that access the local file system will work out of the box with a data grid. This is consistent with our goal of reducing the "activation energy" of grid adoption.

There are three ways for end users to access data in the data grid: via the local file system and an NFS mount of a DGAS, via a set of command lines tools, and via a Web interface. In addition, users may want to share some of their data into the grid, manage access control lists for files and directories that they own, etc.

- NFS—We have already discussed access via the native file system. Applications require no modification, tools such as "ls" in Unix and "dir" in Windows work on mounted data grids. The user is not aware they are even using the ADG. A similar capability using Windows file system access (via CIFS) is also available.

■ Command Line Interface—A data grid can be accessed using a set of command line tools that mimic the Unix file system commands such as *ls*, *cat*, etc. The Avaki analogues are *avaki ls*, *avaki cat*, etc. The Unix-like syntax is intended to mask the complexity of remote data access by presenting familiar semantics to users. The command-line access tools are rarely used by end users and are provided for administrators or for users who may be unable for whatever reason to mount a DGAS into their file system.

■ Web-based Portal—The third access mechanism is via a Web-based portal. Using the portal, a user can traverse the directory structure, manage access control lists for files and directories, and create and remove shares. For example, Figure 8.2 shows the interface to create a new share and map it into the ADG.

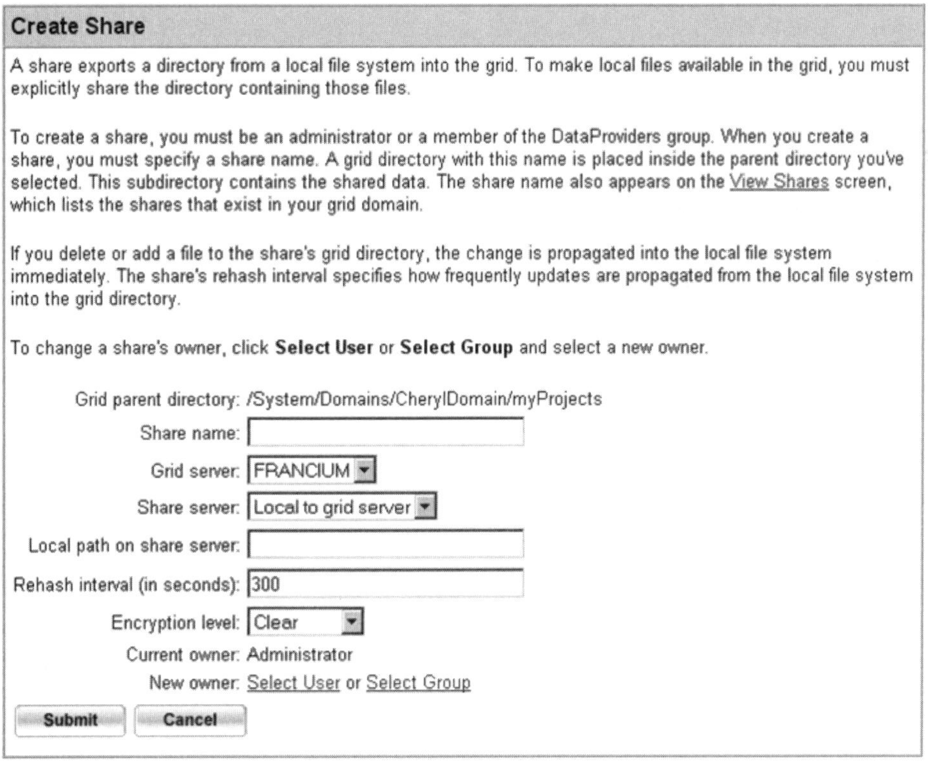

Create Share

A share exports a directory from a local file system into the grid. To make local files available in the grid, you must explicitly share the directory containing those files.

To create a share, you must be an administrator or a member of the DataProviders group. When you create a share, you must specify a share name. A grid directory with this name is placed inside the parent directory you've selected. This subdirectory contains the shared data. The share name also appears on the View Shares screen, which lists the shares that exist in your grid domain.

If you delete or add a file to the share's grid directory, the change is propagated into the local file system immediately. The share's rehash interval specifies how frequently updates are propagated from the local file system into the grid directory.

To change a share's owner, click **Select User** or **Select Group** and select a new owner.

Grid parent directory: /System/Domains/CherylDomain/myProjects
Share name:
Grid server: FRANCIUM
Share server: Local to grid server
Local path on share server:
Rehash interval (in seconds): 300
Encryption level: Clear
Current owner: Administrator
New owner: Select User or Select Group

Submit   Cancel

FIGURE 8.2 The user uses the Web-based GUI to add a share and thus map the data into the global directory structure. The user provides the name of the share, i.e., the path name, the local path of the data, the rehash interval, and the encryption level. In order to share a directory, the user must be a member of the "DataProviders" group. Thus, management may restrict who is allowed to make data available on the grid.

The user can also use the Web-based portal to view and modify access control lists on files, directories, groups, etc. For example, Figure 8.3 shows the ACL for an object "myProjects." Figure 8.4 illustrates changing the ACLs for an object "fred."

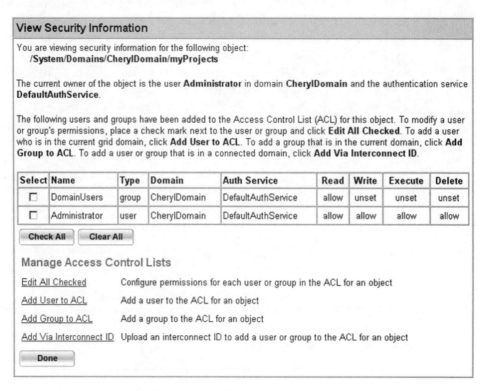

FIGURE 8.3 View/modify access control lists for an object. Both users and groups can be added to the access control lists via the links at the bottom of the page.

8.4.2 Managing the Data Grid

Avaki ensures secure access to resources on the grid. Files on participating computers become part of the grid only when they are *shared*, or explicitly made available to the grid. Further, Avaki's fine-grained access control is used to prevent unauthorized access to shared data. Any subset of objects can be shared—only a specific set of files, for example—and access controls specified for them by data owners. Files or directories that have not been shared are not visible to grid users. By the same token, a user of an individual computer or network that participates in the grid is not automatically a grid user, and does not automatically have access to grid files. Only users who have explicitly been granted access can take advantage of the shared data. Local data owners control access to their data.

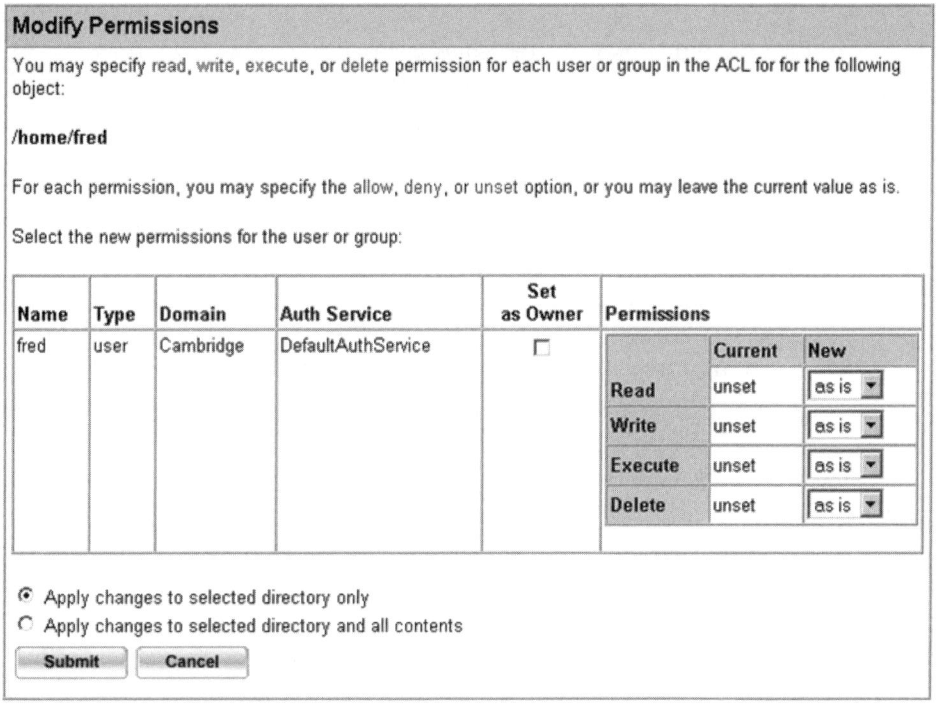

Modify Permissions

You may specify read, write, execute, or delete permission for each user or group in the ACL for for the following object:

/home/fred

For each permission, you may specify the allow, deny, or unset option, or you may leave the current value as is.

Select the new permissions for the user or group:

Name	Type	Domain	Auth Service	Set as Owner	Permissions		
fred	user	Cambridge	DefaultAuthService	☐		Current	New
					Read	unset	as is ▼
					Write	unset	as is ▼
					Execute	unset	as is ▼
					Delete	unset	as is ▼

⦿ Apply changes to selected directory only
◯ Apply changes to selected directory and all contents

[Submit]   [Cancel]

FIGURE 8.4 The access control lists for the object "/home/fred" is being modified. If the object is a directory, the changes can be applied recursively.

The administrative unit of a data grid is the *grid domain*. A data grid can be made up of one grid domain or several grid domains—effectively a "grid of grids." Multiple domains should be established when:

- Networks do not share a namespace. In this case, separate grid domains should be established for those networks, and the two grid domains interconnected.
- The two units truly represent different administrative domains within the organization, or two separate organizations. In these cases, each organization will want to administer its grid separately, and might also be creating grids under different projects with different time constraints.

When different grid domains must share data among them, administrators cooperate to ensure that the appropriate access is granted. The ability to create multiple grid domains and then interconnect them thus allows the organization to leave administration in the hands of local administrators, leave data access control in the hands of local data owners, and still achieve the important benefits of sharing data over a wide area.

Systems administrators perform basic tasks as shown in the systems administrator's main menu Web page (Figure 8.5):

- Server management, where the number of grid servers is specified and hot spares for high availability are configured.
- Grid user management, where users and groups are either imported from the existing LDAP, Active Directory, or NIS environment, or they are defined within the grid itself.
- Grid object management, where files and directories can be created and destroyed, ACL set, and new shares added.
- Grid monitoring, where logging levels, event triggers, and so on are set. ADG can be configured to generate SNMP traps and thus be integrated into the existing network management infrastructure.
- Grid interconnects, where the system administrator can establish and support connections with other grid domains.

In the following sections we will examine in more detail the architecture and servers that sit behind this interface.

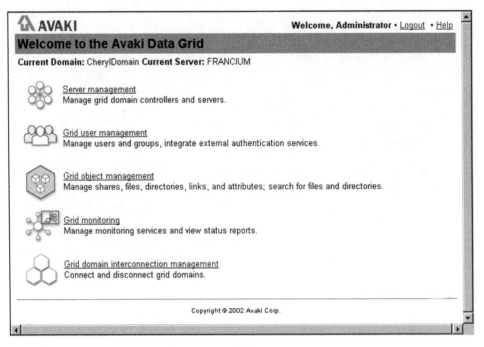

FIGURE 8.5 The main menu for system administrators.

8.5 DATA GRID ARCHITECTURE

ADG 3.0 has been written almost entirely in Java, with a small amount of native code primarily for performing operations not supported by the Java.io package. The architecture is based on an off-the-shelf J2EE application server.

Every ADG component runs within an application server. A Java application server is the equivalent of a traditional operating system, but for J2EE components. Objects are created, deactivated, reactivated, and destroyed within the application server on demand. Interactions between objects within the same application server are processed by the Java Virtual Machine (JVM) within the application server. Interactions between objects in different application servers (typically, on different machines) are processed using remote method invocations (RMI). The product is configured to run RMI using SSL sockets in order to protect user credentials. Interactions between objects may cause their internal states to be changed. The persistent state of objects is stored in an embedded relational database accessed by the application server. All objects log several levels of messages using log4j, which stores these logs in files associated with the application server but can also be configured to generate Simple Network Management Protocol (SNMP) traps, send e-mail, etc.

The major components of an ADG are: grid servers, share servers, data grid access servers (DGAS), and proxy servers. A grid server performs grid operations such as authentication, access control, and meta-data management. A share server performs bulk data transfer between a local disk on a machine and the grid. A data grid access server enables presenting the data grid as a Unix directory or a Windows drive using the NFS protocol. In this section, we discuss the components of an ADG. The interaction between these components provides an insight into the workings of a data grid.

8.5.1 Grid Servers

A *grid server* is the primary component of a data grid. A grid server performs grid-related tasks such as domain creation, authentication, access control, meta-data management, monitoring, searching, etc. When deploying a data grid, the first grid server deployed typically bears the responsibility of starting a grid. This grid server is also called a grid domain controller (GDC). The GDC creates and defines a domain. A domain represents a single grid. Every domain has exactly one GDC. Multiple domains may be interconnected by invoking the appropriate functions on their respective GDCs.

A GDC is sufficient for creating, using, and maintaining a small grid. However, starting multiple grid servers enables scalability for larger grids, as each grid server can host a portion of the grid directory. These grid servers are connected to the

GDC. Each of these grid servers can be made responsible for a subset of the grid-related activities. For example, one of these grid servers can be made responsible for authentication. When a user logs into the grid, this grid server would be responsible for receiving the user name and password and either verifying the user's identity using an in-built grid authentication service or delegating this process to a third-party authentication service such as NIS, Active Directory, or Netegrity. Once the user's identity has been verified, the grid server is responsible for generating credentials for this session for this user.

Another important task performed by a grid server is access control. When a user requests access to a file, directory, or any other object in a grid, the grid server uses her credentials to retrieve her identity and then check the access controls on the object to determine if the requested access is permissible or not. Because the user may issue multiple requests when accessing a large file or when accessing a file repeatedly, the grid server may pass on the permission information in a handle to a share server in order to avoid repeated access control checks.

Yet another important task performed by a grid server is meta-data management. Every object in a grid has meta-data associated with it, such as creation time, ownership information, modification time, etc. Administrators and grid users can also create their own meta-data. A grid server is responsible for storing this meta-data in an internal database, performing searches on it when requested, and re-hashing the information when it becomes stale.

A grid server can also be configured to perform monitoring services on other grid components. Monitoring typically involves determining the response time of other components to ping messages. As the Avaki Data Grid software evolves to incorporate access to relational data, the grid server is expected to perform database tasks such as opening a connection, issuing a query, or executing a stored procedure and reporting results into the data grid.

8.5.2 Share Servers

A *share server* is an ADG component that is responsible for bulk data transfer to and from a local disk on a machine. A share server is always associated with a grid server. The grid server is responsible for verifying whether a given read/write request is permissible or not. If the request is permitted, the grid server passes a handle to the user as well as the share server. The user's request is then forwarded to the share server along with this handle. Subsequent requests are satisfied by the share server without the intervention of the grid server. Naturally, if the user issues a new request, for instance, to a new file, the grid server verifies the request anew before delegating the transfer to the share server.

A share server's main responsibility is to translate a grid read/write request into an equivalent read/write on the underlying file system. Depending on how the share server is configured, the translation may require decrypting data before writing to the file system and encrypting data after it has been read from the file system. Another responsibility of the share server is processing a rehash request initiated by its grid server. A rehash ensures consistency between the grid server's internal database about the contents of a shared directory and the actual contents of the equivalent directory on the file system. Since sharing a directory does not preclude accessing the same directory using OS tools on that machine, it is possible for the contents of a shared directory to be changed without any Avaki component being involved. A rehash restores the consistency of the data grid in such situations. Rehashes may be explicit or periodic.

A share server performs the actual bulk data transfers, whereas its grid server performs grid-related tasks associated with the transfers.

8.5.3 Data Grid Access Servers (DGAS)

A *DGAS* provides a standards-based mechanism to access a data grid. A DGAS is a server that responds to NFS 2.0/3.0 protocols and interacts with other data grid components. When an NFS client on a machine mounts a DGAS, it effectively mounts the entire data grid in a single step, mapping the ADG global name space into the local file system and providing completely transparent access to data throughout the grid without even installing Avaki software. This NFS-based access to an ADG complements the command-line and Web-based access that Avaki provides as part of every data grid deployment. An upcoming version of the DGAS will support the Common Internet File System (CIFS) protocol for Windows clients as well.

Despite the functional similarity, a DGAS is not a typical NFS server. First, it has no actual disk or file system behind it; it interacts with components that may be distributed, be owned by multiple organizations, be behind firewalls, etc. Second, a DGAS supports the Avaki security mechanisms; access control is via signed credentials, and interactions with the data grid can be encrypted. Third, a DGAS caches data aggressively, using configurable local memory and disk caches to avoid wide-area network access. Furthermore, a DGAS can be modified to exploit semantic data that can be carried in the meta-data of a file object, such as "cacheable," "cacheable until," or "coherence window size." In effect, a DGAS provides a highly secure, wide-area NFS.

To avoid the rather obvious hotspot of a single DGAS at each site, Avaki encourages deploying more than one DGAS per site. There are two extremes: one DGAS per site, and one DGAS per machine. Besides the obvious tradeoff between scalability and the shared cache effects of these two extremes, an added security benefit of having one DGAS per machine is that the DGAS can be configured to accept requests from only the local machine, eliminating the classic NFS security attacks via network spoofing.

8.5.4 Proxy Servers

A *proxy server* enables accesses across a firewall. A proxy server requires a single port in the firewall to be opened for TCP—specifically HTTP/HTTPS—traffic. All Avaki traffic passes through this port. Opening a firewall port essentially involves permitting traffic in and out of that port on the firewall machine and forwarding incoming traffic to another machine inside the firewall on which the Avaki proxy server is started. The proxy server accepts all Avaki traffic forwarded from the firewall and redirects the traffic to the appropriate components running on machines within the firewall. The responses of these machines are sent back to the proxy server, which forwards this traffic to the appropriate destination through the open port on the firewall.

A proxy server is associated with a grid domain "inside" a firewall. In other words, the proxy server and other grid servers and share servers must be in a common DNS domain and should be able to send messages to one another freely. Machines "outside" the firewall—i.e., in other DNS domains that are restricted by the firewall—must communicate with machines inside the firewall via the proxy server alone. The machines outside the firewall are not considered part of the grid domain inside the firewall. Therefore, data access through a firewall necessarily involves multiple grid domains connected to one another. Multiple grid domains may access a domain inside a firewall through the same proxy server. Two grid domains that are inside different firewalls may communicate with each other through one proxy server associated with each domain.

A proxy server may encrypt/decrypt as well as compress/uncompress data flowing through it. Message encryption maintains privacy and integrity of data grid traffic, whereas compression reduces network traffic, improving bandwidth. These operations occur transparently from the user's perspective as well as independent of the working of the rest of the grid components.

8.5.5 Failover Servers (Secondary Grid Domain Controllers)

A *failover server* is a grid server that serves as a backup for the GDC. A failover server is configured to synchronize its internal database periodically with a GDC. As a result, if a GDC becomes unavailable either because the machine on which it is running is down or because the network is partitioned or for any other reason, users can continue to access grid data without significant interruption in service.

Grid objects access one another using a unique name, called a Location-independent Object IDentifier (LOID). The Avaki runtime system resolves LOIDs into location-specific identifiers. When a failover server is added to a grid, the address of the failover server is added to the location dependent name of objects that live on the GDC. The run-time communication protocol for every object tries the addresses in the location-specific id in order every time. If the first address—i.e., the GDC address—is unreachable, the object automatically fails over to the next address—i.e., the failover server. If even that address is unreachable, the object reports an error and terminates the action.

The database within the failover server is synchronized with the database of the GDC periodically. If a GDC becomes unavailable, the database in the failover server is guaranteed to be closely consistent with that of the GDC. However, subsequent actions may make the failover server database inconsistent with that of the currently unavailable GDC. Therefore, when a data grid is operating in failover mode, i.e., with a failover server acting in lieu of a GDC, actions that change the GDC's database are prohibited. Typically, this prohibition means that adding new shares to the GDC or adding new files and directories to existing shares on that GDC is prohibited. This solution avoids some of the more difficult problems of fault-tolerance on a grid. For example, after a GDC becomes unavailable, the remaining grid servers do not have to vote among themselves to pick a new GDC—the next failover server listed in location-specific ID automatically acts as a limited GDC. When the GDC returns, again no voting is required to pick the primary component of the grid—the GDC plays that role simply because it continues to be the first address in every location-specific name.

8.6 CONCLUSION

In this chapter, we have presented Avaki Data Grid software and the data grids created by installing this software. We have discussed use of the data grid, the architecture, and alternatives to data grid technology—ftp, scp, wide area NFS, and DMZs and found that these alternatives fall short on several dimensions.

Data grid software, such as the Avaki Data Grid, provides high-performance; easy, transparent, secure collaboration; and coherent sharing between different administrative domains and organizations. This allows organizations to reduce the "friction" of collaboration both internally and externally, reducing both costs and time to market.

Acknowledgments

Many people contribute to a project as large as Avaki Data Grid. Without them this chapter would not be possible. The engineering team includes (in alphabetical order) Josh Apgar, Steph Bacon, Eli Daniel, Peter Fein, Michael Herrick, John Karpovich, Duane Merrill, Mark Morgan, Anh Nguyen-Tuong, Rudi Seitz, and William Tam. I would also like to specially thank Anand Natrajan for his many contributions, and Linda Thorsen.

9

The Open Grid Services Architecture

Chris Smith and Ian Lumb

Platform Computing, Inc.

In This Chapter

- Open Grid Services Architecture (OGSA) Defined
- Open Grid Services Infrastructure (OGSI) Defined
- Building on OGSA Platform

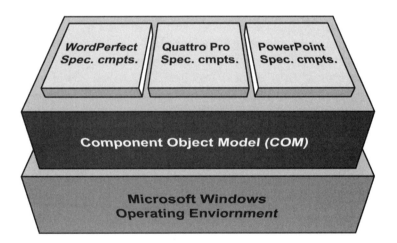

9.1 INTRODUCTION

The increasing adoption of highly distributed virtual environments requires a change in how technology is developed, implemented, and supported. As with any new computing model, standards are vitally important to pave the way for interoperability, to provide a standard set of interfaces for vendors to develop applications against, and to provide a standard interface to underlying computing resources. Enterprise customers are bullish on standards compliance and demand that vendors stand behind these claims. The future of Grid Computing is dependent on a common set of standards that provide the collaborative context for partners to work together and for vendors to provide customers with interoperable solutions.

Today, the most widely accepted grid standards are driven out of the Global Grid Forum (GGF) (see Chapter 4), a community-initiated forum of over 5000 individual researchers and practitioners working on distributed computing, or "grid" technologies. GGF's primary objective is to promote and support the development, deployment, and implementation of grid technologies and applications through the creation and documentation of "best practices"—technical specifications, user experiences, and implementation guidelines.

The Open Grid Services Architecture (OGSA) (see Chapter 4) is a set of technical specifications which define a common framework that will allow businesses to build grids both across the enterprise and with their business partners. It is expected that OGSA will define the standards required for both open source and commercial software for a broadly applicable and widely adopted global grid infrastructure.

The Open Grid Services Architecture (OGSA) has been proposed as an enabling infrastructure for systems and applications that require the integration and management of service within distributed, heterogeneous, dynamic "virtual organizations." [Physiology][1]

This enabling infrastructure defines the notion of a "Grid Service," which is a Web Service that conforms to a specific interface and behavior, as defined in various specifications developed by the Global Grid Forum (GGF). Moreover, the OGSA defines a set of Grid Services that can be used to implement various application patterns evident in distributed computing.

Although previously the domain of academia and scientific computing, Grid Computing is now being adopted by mainstream industry as a mechanism for allowing organizations to more effectively utilize and manage their IT infrastruc-

---

[1] [Physiology] The Physiology of the Grid: An Open Grid Services Architecture for Distributed Systems Integration, I. Foster, C. Kesselman, J. Nick, S. Tuecke, Authors. Globus Project, 2002. Available at *http://www.globus.org/research/papers/ogsa.pdf*

tures. As the scale of these grid deployments start crossing geographies and different organizational domains (perhaps within partner or service grids), the need to integrate heterogeneous environments in a cost effective way becomes more evident. Companies developing grid solutions today see OGSA compliance and OGSI implementations as key enabling technology of the future, increasing the adoption of grid solutions within industry as a whole, resulting in benefits to commercial customers looking for the scalable, integrated solutions that grid technologies can help to provide.

This chapter is intended to present an overview of the OGSA Platform and related specifications. Through examination of the specifications available from the GGF pertaining to OGSA, the various parts of OGSA will be described, with a view to summarizing the plethora of information and interpretations that exist surrounding this guideline. Furthermore, demonstrating the broad support for OGSA, some specifications based on OGSA will be described, and an overview of some current implementations of technology based on OGSA.

9.2 AN ANALOGY FOR OGSA

To fully appreciate OGSA, it is helpful to start with an analogy. The analogy provides before-and-after reference points through a familiar example. Although there are many potential analogies, office-productivity software is used here. Because the analogy does not provide a completely accurate representation, it is introduced here, and revisited later in this chapter.

Figure 9.1 illustrates office-productivity software in the earliest releases of the Microsoft Windows operating environment: Multiple vendors with multiple products. Despite the common operating environment, vendors had to develop their own user interfaces, build their own tools and utilities, etc. End users had to learn several different user interfaces, plus the details of per-application spell checkers, print drivers, etc. With each product an island unto itself, literally adrift in the operating environment, it soon became evident that this approach was not sustainable.

With the release of version 3.1 of the Microsoft Windows Software Development Kit (SDK), an architectural paradigm shift was introduced (Figure 9.2). Key to this shift was enabling technology known as Object Linking and Embedding (OLE).

OLE is a technology that enables an application to create compound documents that contain information from a number of different sources. For example, a document in an OLE-enabled word processor can accept an embedded spreadsheet

FIGURE 9.1 Multiple vendors with multiple products sharing a common operating environment.

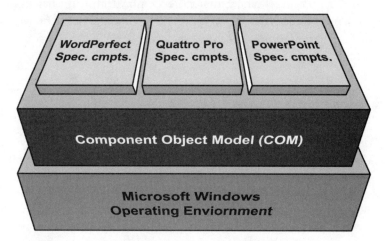

FIGURE 9.2 Multiple vendors with multiple products under the COM-enhanced Microsoft Windows operating environment.

object. Unlike traditional "cut and paste" methods where the receiving application changes the format of the pasted information, embedded documents retain all their original properties. If the user decides to edit the embedded data, Windows activates the originating application and loads the embedded document. [OLE][2]

OLE has been superceded by the Component Object Model (COM)—Microsoft's framework for developing and supporting program component objects.[COM][3] The COM-enhanced Microsoft Windows operating environment permits isolated products to be focused on points of differentiation, while leveraging a common base.

---

[2] [OLE] *http://support.microsoft.com/support/kb/articles/Q86/0/08.asp*

[3] [COM] *http://www.microsoft.com/com*

The introduction of COM:

- Introduces a de facto standard and implementation—encouraging modularity and facilitating extensibility in the Microsoft Windows operating environment.
- Facilitates the transition of isolated products to a suite of integrated products—almost eliminating the value of per-application differentiation, and favoring a single-vendor solution.
- Enables customers—from end user to support organizations.
- Enables independent software vendors (ISVs)—leveraging COM simplifies and accelerates software development.

COM has closer technical ties with OGSA than this analogy might indicate. DCOM, the distributed version of COM, is generally equivalent to CORBA (Common Object Request Broker Architecture). Although OGSA leverages CORBA concepts, OGSA also needs to directly address secure interoperability, and provide a richer interface definition language. The path to OGSA is considered in the following section.

9.3 THE EVOLUTION TO OGSA

This section discusses the historical evolution of the OGSA standard.

9.3.1 Grid Computing

In Anatomy,[4] the "Grid problem" is defined "as flexible, secure, coordinated resource sharing among dynamic collections of individuals, institutions, and resources—what we refer to as virtual organizations." The purpose of such sharing is to enable a new kind of eScience, by harnessing collections of resources that could not be "brought under one roof." Existing distributed computing technologies such as CORBA or DCE, although effective within the domain of one organization, suffered from the fact that often the relationships between organizations were statically defined, making the management of such relationships unscalable when the number of organizations needing to collaborate grew larger. With many requirements coming from the field of high performance computing, grid technologies such as the Globus Toolkit were developed to help solve this grid problem. Primarily deployed within research communities, where the sharing of resources was often crossing

---

[4] [Anatomy] The Anatomy of the Grid: Enabling Scalable Virtual Organizations, I. Foster, C. Kesselman, S. Tuecke, Authors. International Journal of High Performance Computing Applications, 15 (3). 200–222. 2001. Available at *http://www.globus.org/research/papers/anatomy.pdf*

organizational boundaries, it was quickly recognized that there was a need for a common set of APIs, protocols, and services to support Grid Computing in a way that would facilitate the dynamic interoperation required by the virtual organization.

Growing out of a number of discussions, meetings, and Birds of a Feather concerned with grid technologies (*http://www.ggf.org/Meetings/history/history.htm*), the Global Grid Forum was formed to host a number of working groups focused on defining standards and best practices for distributed computing. There are focus areas covering specific aspects of the grid problem, which include working groups on security issues, data management and access, resource management and scheduling, and user environments and programming models. Again, many of the requirements drew on the experiences of academia, and the main focus was firmly on high performance computing.

Meanwhile, aspects of the grid problem appeared outside of the halls of academia. As modeling and simulation applications became more sophisticated, driving industrial manufacturing and research to use more computational resources, various companies, including Platform Computing, were solving subsets of the grid problem in order to enable these organizations to realize more efficient utilization of computing resources, thus helping companies achieve ROI on computing investments. Witness the increased involvement of software and system vendors in the GGF, who see the benefits of applying grid technologies to solve business related issues.

9.3.2 Web Services

In parallel with the development of grid technologies for scientific computing, businesses were attempting to deal with data and application integration problems as businesses began relying more on IT. Requirements for B2B ecommerce, partner networks, and internal application integration were consistently hitting up against constraints of many existing distributed computing models, many of which attempted to be overarching and complete. A business based on CORBA couldn't easily integrate systems with businesses based on Java RMI, since there were incompatibilities at the protocol level and at the interface level.

In response to these kinds of problems, Web Services were defined to decouple the programming environment from the integration environment. Based on existing Internet standards such as XML, W3C standards such as Simple Object Access Protocol (SOAP), Web Services Description Language (WSDL), and Web Services Inspection Language (WSIL) help define conventions for a service provider and a service consumer to exchange messages in a protocol and programming language independent way. Furthermore, rather than being yet another limited deployment messaging mechanism, the industry adoption of the Web Services model has exploded. For example, Microsoft (.NET™), IBM (Dynamic eBusiness™), and Sun (Sun ONE™) all base their technologies on Web Services. Because Web Services are programming language and operating environment neutral, developers can leverage existing programming skills, and legacy applications can continue to be maintained, thus accelerating system integration, as the focus need only be on the interaction between systems and the semantics of the messages being sent.

The ability to rapidly integrate applications using a service-oriented model also allowed organizations to take advantage of new paradigms in computing architecture, such as cluster computing. Applications which previously needed to run on large, monolithic systems could now be architected as sets of services and be run on more cost effective clusters of Intel-based systems, while still maintaining previous levels of QoS.

9.3.3 Convergence

The utility of Grid Computing was also being recognized in commercial computing environments where the application integration problems were starting to look a lot like the grid problem, with issues around security, data management, and effective use of computing resources.

From Physiology:

Nevertheless, we argue that grid concepts are critically important for commercial computing not primarily as a means of enhancing capability, but rather as a solution to new challenges relating to the construction of reliable, scalable, and secure distributed systems. These challenges derive from the current rush, driven by technology trends and commercial pressures, to decompose and distribute through the network previously monolithic host-centric services.[1]

Again in Physiology, the case is made that the Web Services approach provides a number of desirable characteristics as a grid technology. The focus on interface rather than implementation, as well as the need to support activities such as dynamic discovery of available services and interoperation of heterogeneous environments, provides the necessary level of abstraction to facilitate the dynamic relationships represented by virtual organizations. In addition to the technical benefits, adopting Web Services mechanisms allows for the accelerated adoption of grid technologies based on Web Services due to the large number of tools, services, and hosting environments that are already in widespread use.

The Open Grid Services Architecture (OGSA) is being proposed to the GGF as a standard platform for Grid Computing, and represents the definition of grid technology and functionality as Web Services.

9.4 OGSA OVERVIEW

The OGSA is intended to define an infrastructure for integrating and managing services within a virtual organization. More simply, it is an architecture for building grid applications. The concept of OGSA is introduced in Physiology, but is being defined more strictly in OGSA Platform[5]. The specification is being developed under the OGSA Working Group of the GGF (*https://forge.gridforum.org/projects/ogsa-wg*). The scope of the working group is described in OGSA Platform and includes identifying and outlining requirements for the different OGSA services, protocols, and usage models, as well as defining relationships between OGSA and other standards activities (e.g., from Organization for Advancement of Structured Information Standards (OASIS) or W3C). The goal is to define standard approaches to solving basic problems that are common to many grid deployments.

The approach of the working group is to drive this work based on use case scenarios, some of which include:

- Commercial Data Center—management of outsourced computing infrastructure
- National Fusion Collaboratory—on-demand application services for large-scale analysis and simulation
- Severe Storm Prediction—comparison of models to "streamed" sensor data triggered by a "storm event"
- Online Media and Entertainment—video on demand or online gaming
- Service-Based Distributed Query Processing—data and computing resource federation

The analysis of these use cases and of other both GGF and non-GGF documents helps identify common patterns of activity, which can indicate what types of functional requirements are needed to support grid environments in general. The functional requirements fall into different categories:

---

[5] [OGSA Platform] *The Open Grid Services Architecture Platform*, I. Foster, D. Gannon, Editors. Global Grid Forum. draft-ggf-ogsa-platform-2.

- Support for heterogeneity—of platforms (OSes, hosting environments, etc.), mechanisms (different security schemes or device access interfaces), and administrative environments (often expressed with different and potentially conflicting policies around system usage).
- Different application structures—single-process and multi-process applications with varying resource requirements, flows of interacting applications to be treated as higher-level tasks, and workload, which is a combination of many varying processes and flows.
- Basic functionality—discovery and resource brokering, metering and accounting, data sharing, support for managing virtual organizations, monitoring, and policy enforcement.
- Security—need to support multiple security infrastructures and perimeter security solutions such as firewalls.
- Resource management—provisioning of resources in a uniform way, virtualizing access to resources, optimization of resource usage (while still meeting QoS constraints), managing transport (i.e., for data sharing), batch and interactive access to resources, SLA management and monitoring, and CPU scavenging.
- Some desirable system properties—fault tolerance, disaster recovery, self-healing capabilities, ability to detect defects or attacks and to route around same, management of legacy applications, and administration capabilities (for automating common management functions).

Although not comprehensive, this extraction of functional requirements from the analysis of use cases helps guide the development of the components of the OGSA Platform. OGSA Platform will continue to be refined as more use cases are developed and new functional requirements are identified.

9.4.1 The OGSA Platform

The OGSA Platform is made up of three components: the Open Grid Services Infrastructure (OGSI), the OGSA Platform Interfaces, and the OGSA Platform Models.

OGSI represents the convergence of Web Services and grid technologies. It defines the underlying mechanisms for managing Grid Service instances (e.g., messaging, lifecycle management, etc.).

OGSA Platform Interfaces are OGSI-compliant Grid Services (i.e., interfaces and associated behaviors) that are not defined within OGSI. The focus here is on defining the higher level—but basic—services common in many grid deployments. Examples include registries, data access and integration, resource manager interfaces, etc.

OGSA Platform Models are the combination of OGSA services and information schemas for representing real entities on the grid. For example, a standard definition of terms describing a computer system and the associated behavior is an example of a model for a computer system.

The OGSA Platform focuses on interfaces and usage models; it does not define protocol bindings, hosting environments, or domain-specific services. These are intended to be defined in other specifications and are referred to as OGSA Platform Profiles.

9.4.2 OGSI

From OGSI: "The OGSI defines mechanisms for creating, managing, and exchanging information among entities called Grid Services." Everything in OGSA is a Grid Service —from registries, to computational tasks, to a data acquisition device, to an SLA—so Grid Services can be permanent or transient, long or short lived. Each Grid Service is required to support a specific set of interfaces, and to act in a specifically defined way, so that two services (even if not "pre-programmed" to work together) can communicate in a meaningful way and act in a predictable fashion.[6]

In order to address some of the shortcomings of the current Web Services specifications, OGSI extends WSDL and XML Schema Definition to support:

- Stateful Web Services
- Inheritance of Web Services interfaces (`portTypes`)
- Asynchronous notification of state change
- References to instances of services
- Collections of service instances
- Service state data that augment the constraint capabilities of XML Schema Definition

All these elements will be described at further length later.

OGSI provides the specifications for how clients deal with Grid Services. It specifies naming and referencing of Grid Services, basic interfaces that all Grid Services must implement in order to be OGSI-compliant, and optional interfaces that can be used to implement common behaviors such as creating services or generating events.

OGSI does not define how Grid Services are implemented within a particular hosting environment. This means that while it might not be possible to take a Grid Service coded in one environment and port it to another, a client that follows the

---

[6] [OGSI] *Open Grid Services Infrastructure (OGSI) Version 1.0*, S. Tuecke, K. Czajkowski, I. Foster, J. Frey, S. Graham, C. Kesselman, T. Maquire, T. Sandholm, D. Snelling, P. Vanderbilt, Editors. Global Grid Forum. draft-ggf-ogsi-gridservice-29.

defined Grid Service conventions can interact with this Grid Service. This is a powerful concept. Clients can be written once to use a particular type of Grid Service, and can interact with any service of that type no matter how or where it is implemented or hosted. The Grid Service can be virtualized away from the operating system or hosting environment, so that services can be upgraded or ported to new environments as required without needing to change the clients. Notice that this distinguishes OGSI from a distributed object model, where development interfaces and hosting environments are fixed. So, if an organization chooses to implement an infrastructure based on Java RMI, the client needs to speak RMI, and the server needs to be built in a Java environment. Generally, this locks developers into particular architectures and programming languages. With the OGSI (i.e., Web Services) approach, the clients and servers can be written in any programming language that supports the Web Services model (e.g., SOAP over HTTP), facilitating change to another programming model and environment when deemed necessary or beneficial. Client programming models are discussed at further length in [OGSI].[7]

OGSI definitions can be grouped into four broad categories: WSDL extension and conventions, service data, core grid service properties, and portTypes for basic services.

9.4.2.1 WSDL Extensions and Conventions

OGSI extended WSDL 1.1 to correct two deficiencies. First, there was no interface inheritance (i.e., inheritance of portTypes). Second, it wasn't possible to add informational elements to the portType (i.e., the set of XML elements allowed in the portType element was rigidly controlled). Although these deficiencies are addressed in WSDL 1.2, at the time of the writing of OGSI these deficiencies needed correction.

Thus, OGSI redefines the wsdl:portType element to add the desired semantics. A separate namespace with the prefix gwsdl was defined in order to isolate all modifications to WSDL 1.1. It is the intention of the GGF to update OGSI when WSDL 1.2 is published as a draft specification.

9.4.2.2 Service Data

In order to allow for stateful Web Services, OGSI defined a mechanism to expose a service instance's state data, called serviceData. The concept is roughly equivalent to declaring attributes in an object-oriented interface definition language. These data can be used for reading, writing, or subscription. Through the use of a set of

---

[7] [OGSI] *Open Grid Services Infrastructure (OGSI) Version 1.0*, S. Tuecke, K. Czajkowski, I. Foster, J. Frey, S. Graham, C. Kesselman, T. Maquire, T. Sandholm, D. Snelling, P. Vanderbilt, Editors. Global Grid Forum. draft-ggf-ogsi-gridservice-29.

new XML elements, serviceData extends WSDL so that in addition to defining the operations available to be invoked, arbitrary state information can also be made publicly accessible. Furthermore, the base GridService portType (from which all Grid Services inherit), defines operations for accessing this serviceData by name.

An important attribute of the serviceData element (SDE) is its mutability. This is an indication of how the SDE value can change over time, and is thus important for clients needing to understand the state changes of the Grid Service being accessed. Mutability can take one of four values:

- Static—the SDE value is assigned in the WSDL service definition. This is a similar notion to a "class member variable" in object oriented programming languages.
- Constant—the SDE value is assigned when the Grid Service is instantiated and must not change during the lifetime of the service.
- Extendable—this value implies that once elements are added to the SDE value (note that SDEs are allowed to have multiple elements), they are part of the SDE for the lifetime of the Grid Service. New elements can be added, but none are removed.
- Mutable—any elements of the SDE may be added or removed at any time.

Each service has a collection of serviceData elements from all the portTypes the service inherits. The service must have a "logical" XML document containing these values, with a root element type of serviceDataValues. The service implementation is free to store these values in any fashion, but must be able to convert the internal representation to XML as necessary.

9.4.2.3 Core Grid Service Properties

A section of OGSI describes some properties common to all Grid Services. These properties are subsequently used in the definition of OGSI Grid Service interfaces and help define the semantics of the Grid Service.

- Service Description and Service Instance—OGSI makes a point of distinguishing between the description of a Grid Service and instances which implement the Grid Service. The Grid Service description specifies how clients interact with instances of the service. The description is not dependent on any particular instance. The Grid Service description can also be used for discovering instances of Grid Services that implement a particular service description, or to find factories that can create specific service types. The service description is also intended to capture some semantics about the Grid Service. Syntax is described using WSDL, but there is an intention to allow semantics to be inferred through the name of the portType. Semantics should be attached to the names in specification documents, and perhaps by using Semantic Web descriptions in the future.

- Modeling Time in OGSI—The Greenwich Mean Time (GMT) global time standard is used so that times can be referred to unambiguously, but this does not imply any kind of time synchronization between Grid Services. Although service hosting environments are encouraged to use mechanism such as Network Time Protocol (NTP) to synchronize clocks, this is not required, so clients and services cannot make any assumptions that clocks are in synch. Of course, some specific OGSI-compliant services might specify requirements for time synchronization above and beyond OGSI. Kerberos services come to mind as a service that requires clock synchronization between client and server. OGSI also defines conventions for representing "zero time" and "infinity."

- XML Element Lifetime Declaration Properties—Because SDEs generally represent snapshots in time, there is a need to be able to describe the lifetime that a particular SDE value is valid for. Three XML attributes are defined which can be attached to any XML element that allows for extensibility attributes, including SDEs. These attributes are very useful in the context of SDEs where they can be used by clients to help decide how to set state polling intervals, etc. The three attributes are:

  - ogsi:goodFrom: the time from which the content of an element is valid.

  - ogsi:goodUntil: the time when the content can be considered invalid. Must be greater than or equal to goodFrom.

  - ogsi:availableUntil: the time up until the element itself can be considered valid. Before availableUntil, a client should be able to read the element value again, but after availableUntil, this element might no longer exist.

- Interface Naming and Change Management—Grid Service semantics are characterized by the interface specification (i.e., its portTypes) and the implementation of this interface within a particular hosting environment. In order to ensure that clients get reliable and repeatable semantics, there need to be mechanisms for clients to discover if a service has changed in a way that is not backward compatible (e.g., the number of arguments for an operation changes). To accomplish this, OGSI specifies a naming scheme for portTypes to make sure that changes are detectable.

  From OGSI: In OGSI, our concern with change management leads us to require that all elements of a Grid Service description MUST be immutable. The Qname[8] of a Grid Service portType, operation, message, serviceData declaration

---

[8] A Qname is an XML element name which is "fully qualified" using both the namespace and the local name of the element together, thus unambiguously identifying the XML element.

and underlying type definitions MAY be assumed to refer to one and only one WSDL/XSD specification. If a change is needed, a new portType MUST be defined with a new QName—that is, defined with a new local name, and/or in a new namespace.

In short, any changes to the service require that the service get a new name, so clients will not implicitly access new implementations unwittingly.

■ Naming Grid Service Instances—OGSI defines a two-level naming scheme for locating Grid Services. One or more Grid Service Handles (GSH) names every Grid Service. The GSH names only one Grid Service instance, globally and for all time. The GSH itself is not enough to access the actual Grid Service. It is resolved (through a service which implements the HandleResolver portType) into a Grid Service Reference (GSR). Although the GSH is protocol and binding independent, the GSR is specific to the binding mechanism used by a client to access the service. So, for example, if SOAP is used to access the service, the GSR would be a properly formed WSDL document. [OGSI][6] goes into much more detail about the encoding and semantics of GSHs and GSRs. Another concept introduced is the service locator. Simply, the locator is an XML structure containing zero or more GSHs, zero or more GSRs, and zero of more interface (i.e., portType) QNames. All the GSHs and GSRs must point to the same Grid Service instance, and the interfaces listed must be implemented by the Grid Service instance.

■ Grid Service Lifecycle—The life of a Grid Service instance is demarcated by its creation and destruction. Although the mechanism for performing this action is hosting environment-specific, clients of a service need to be able to interact with these capabilities. Clients will generally create Grid Services by invoking the createService operation on a Grid Service that implements the Factory portType. Destruction, on the other hand, can be accomplished in a couple of ways. First, the client can explicitly destroy the Grid Service. Second is a "soft-state" approach where clients indicate their interest in a Grid Service for a particular duration of time. After that time, the service may be automatically destroyed. Clients also can reaffirm interest and extend this termination time if necessary.

■ Common Handling of Operation Faults—OGSI defines an XML base type (ogsi:FaultType) which must be returned in all fault messages from Grid Services. Information defined in this base type includes a description (human readable), an originator (using a service locator), a timestamp, a fault cause (which is itself an ogsi:FaultType, and is used for more specific fault description), a fault code (which can be used to map legacy error codes such as POSIX errno), and an extension element for providing any additional information. Faults are defined in the WSDL for the Grid Service.

■ Extensible Operations—Many OGSI operations can accept an "untyped" input argument, which allows common patterns of behavior to be expressed, without needing to define an operation for every different type that can be used in the pattern. For example, the `NotificationSource::subscribe` operation allows a client to ask for notification messages to be sent if some subset of the services instances's `serviceDataValues` change. The `serviceDataValues` of interest are indicated by a "subscription expression" which is an extensible argument; thus a service inheriting the `NotificationSource` interface can implement customized query mechanism that may have more capability—or perhaps be more domain specific—than the default query mechanism. Extensible parameters are defined using `serviceData` within a `portType`'s operations. The extensible parameter is defined in a `serviceData` element of type `ogsi:OperationExtensibilityType`, and the valid types for the operation are listed in `staticServiceDataValues`.

9.4.2.4 `PortTypes` for Basic Services

OGSI defines a set of `portTypes` and describes the behavior of a collection of common distributed computing patterns that are fundamental to OGSI. The `portTypes` defined (in the ogsi namespace) as taken from [OGSI] [6] are:

- **`GridService`**—encapsulates the root behavior of the service model.
- **`HandleResolver`**—mapping from a GSH to a GSR.
- **`NotificationSource`**—allows clients to subscribe to notification messages.
- **`NotificationSubscription`**—defines the relationship between a single `NotificationSource` and `NotificationSink` pair.
- **`NotificationSink`**—defines a single operation for delivering a notification message to the service instance that implements the operation.
- **`Factory`**—standard operation for creation of Grid Service instances.
- **`ServiceGroup`**—allows clients to maintain groups of services.
- **`ServiceGroupRegistration`**—allows Grid Services to be added and removed from a `ServiceGroup`.
- **`ServiceGroupEntry`**—defines the relationship between a Grid Service and its membership within a `ServiceGroup`.

In order to be OGSI compliant, a Web Service must implement the `GridService` interface. All other `portTypes` are optional. The definitions of the `portTypes` and their behaviors are well described in [OGSI],[9] but it is worth going into detail on the `GridService` interface, in order to see how some of the properties described previously can be applied. `GridService` includes the following `serviceData` elements:

---

[9] [OGSI] Open Grid Services Infrastructure (OGSI) Version 1.0, S. Tuecke, K. Czajkowski, I. Foster, J. Frey, S. Graham, C. Kesselman, T. Maquire, T. Sandholm, D. Snelling, P. Vanderbilt, Editors. Global Grid Forum. draft-ggf-ogsi-gridservice-29.

- interface a list of the QNames of all portTypes implemented by the service.
- serviceDataName—a list of QNames of all SDEs supported by this service instance. This includes SDEs defined at the interface level, as well as SDEs added dynamically during the lifetime of the service instance.
- factoryLocator —a service locator that points to the Grid Service instance that created this Grid Service instance.
- gridServiceHandle—zero or more GSHs of this Grid Service instance.
- gridServiceReference—zero or more GSRs of this Grid Service instance.
- findServiceDataExtensibility—a set of operation extensibility declarations for the findServiceData operation. The client can use a query expression that conforms to any of the listed inputElement types.
- setServiceDataExtensibility—operation extensibility declarations for the setServiceData operation. Similar to findServiceDataExtensibility.
- terminationTime—the termination time for the service.

There are also three staticServiceDataValues elements defined in the GridService portType, which correspond to allowable inputElement types for the findServiceData and setServiceData operations. The operations defined for GridService are as follows:

- findServiceData—query the service data. Takes a QueryExpression as input, and returns a result that is formatted based on the type of QueryExpression (which is an extensible type). Every Grid Service must support a QueryExpression type corresponding to the queryByServiceDataNames parameter as defined in OGSI. Basically, every Grid Service should allow a client to query serviceData values based on the name of the SDE of interest.
- setServiceData—allow for the modification of an SDE's values. Input is an UpdateExpression, and the result is dependent on the type of UpdateExpression. Two types must be supported for UpdateExpression: setByServiceDataNames, which allows a client to change the value of an SDE based on the name of the element, and deleteByServiceDataNames, which allows a client to remove all SDE values for the SDEs that correspond to the provided names.
- requestTerminationAfter—ask that the termination time of the service be changed. It takes a termination time as input and returns the new termination time. These are not necessarily the same, as the input argument corresponds to the earliest time acceptable to the client for service termination, but the server might have other constraints which do not allow it to match the time (e.g., another client registered interest also).
- requestTerminationBefore—similar to requestTerminationAfter, except that the client provides the latest desired termination time.

■ `destroy`—explicitly request the destruction of the service. Note that the Grid Service is not required to destroy itself when a client asks, but can return a fault indicating that it will not be destroyed. This might be useful if the client does not have the right "permissions" to destroy a service, or if other clients are also still using the service.

9.4.3 OGSA Platform Interfaces

The OGSA Platform Interfaces (and related behavior) define a number of functions (as Grid Services, of course) that commonly occur within Grid systems. The functions are divided into a number of categories, which are listed and described below. Note that the interfaces (i.e., the `portTypes`) for these services are not necessarily defined as of yet, and in general the specification will be delegated to other specification documents (e.g., the security services are being defined by the OGSA Security WG at the GGF). The defined functions are:

■ Service Groups and Discovery Interfaces—The two-level naming defined by OGSI based on GSHs and GSRs provide a way of accessing a known service, but for a number of reasons, a higher level abstraction is necessary to allow clients to find services. First, the GSH doesn't contain any semantic information, so you can't discover any attributes of a service instance based on this information. Second, one might not have a locator at all, and still need to discover available services, perhaps based on abstract service attributes. Third, the service locator doesn't distinguish between services you are entitled to use, or QoS attributes of a service, etc. For these reasons, registries are often used to attach more semantic content to a group of services. OGSA Platform outlines two approaches to doing registries, one based on attribute naming (i.e., associating meta-data with service listings), and another based on path naming (i.e., organization of services via a hierarchical naming scheme). Finally, one can see how this builds on the OGSI spec, as registries in general will probably inherit from the `ServiceGroup portType`.

■ Service Domain Interfaces—A common usage pattern in grid solutions is to create Grid Service collections that together produce a higher order Grid Service interface. Usually this is done in order to implement a domain-specific solution based on more generic lower level services. These are called "*Service Domains*" in OGSA Platform. @t:This concept of Service Domain generates some requirements for the registration, discovery, selection, filtering, routing, failover, creation, destruction, enumeration, iteration, and topological mapping of service instances within the Service Domain. Building on OGSI service group interfaces, there is a need to define OGSA Platform interfaces to implement behaviors for filtering, selection, topology, enumeration, discovery, and policy.

■ Security—OGSA Platform security mechanisms are primarily focused on integrating and unifying current underlying security systems that are already in use today (e.g., PKI (Public Key Infrastructure) or Kerberos). GGF's OGSA-Sec working group is defining specifications for generic security services, and their mappings to underlying security systems. There is also an effort to stay consistent with the emerging Web Services security framework being defined in other standards organizations (e.g., in OASIS). Given the dynamic nature of virtual organizations, the OGSA Platform security model needs to support the following security disciplines: authentication, confidentiality, message integrity, policy expression and exchange, authorization, delegation, single logon, credential lifespan and renewal, privacy, secure logging, assurance, manageability, firewall traversal, and security at the OGSI level.

■ Policy—Many Grid Services will need some form of policy management in order to help direct their actions. For example, if a user needs to consume a computing resource, there may be limits on how long it can be used, or what times of day the user can access the resource. Policies are needed to guide resource usage as well as incorporate business logic and service level agreements. Higher-level policies will be required to be representable in a canonical way which "makes sense" to the underlying lower-level resource. To this end, the OGSA Platform needs to define representations of policy, and the functions needed to implement a policy management system. These representations and functions include: a canonical representation for expressing policies, a management control point for policy lifecycle, an interface that policy consumers can use to retrieve required policies, a way to express that a service is "policy aware," and a way to effect change on a resource. OGSA Platform defines an architecture for this set of services.

■ Data Management Services—In the grid environment, there is a great variance in the form of data sources, their location relative to consuming services, their types, and their lifetimes. This diversity adds much complexity when trying to integrate multiple data sources for processing or for management. To this end, it would be useful to define data management interfaces that would abstract the details of accessing a particular set of data from its form. Some of the areas that need to be examined include data access, data replication, data caching services, meta-data catalogs (i.e., specialized registries), schema transformation, and storage.

■ Messaging and Queuing—The OGSA Platform is intended to extend the OGSI notification interfaces to support a larger range of semantics for messaging. Semantics might include attributes such as reliability of message transport, ordering semantics, and routing.

■ Events—An event represents a state change in a system that might be of interest to some other party. The OGSA Platform is intended to define standard representations and standard ways of processing and communicating events in order to enhance interoperability. This mechanism could be drawn from the work of the OASIS Management Protocol TC.

■ Distributed Logging—Distributed logging can be viewed as a special case of messaging, where one entity generates "log artifacts," which may not be used until a later time by some message consumer. OGSA logging can leverage the OGSI notification mechanism, but the semantics of log generation and consumption need to be defined. Logging services need to provide extensions to deal with issues such as the decoupling of log generation from log consumption, putting log records into a common format, filtering and aggregating log records, how long the log records should be kept, and various consumption patterns (e.g., realtime vs. historical requirements). To this end, OGSA defines an architecture for OGSA logging services intended to meet these requirements.

■ Metering and Accounting—Although different grid deployments use different combinations of services and generally are motivated by different economic and business factors, it is a fairly universal requirement to be able to account for resource consumption and utilization over time and identify how those resources are being used. This information is useful for the purposes of trend analysis, capacity planning, chargeback, evaluation of SLAs, detecting faults, etc. The OGSA Platform metering and accounting interfaces are intended to support this activity, through the following interfaces:

   ■ Metering Interface—provides a mechanism to aggregate the metered information from lower level resource (e.g., the operating system accounting logs), in order to unify the resource information across the components involved in the grid application.

   ■ Rating Interface—used to translate the metered information into financial terms (e.g., for the purposes of chargeback).

   ■ Accounting Interface—helps manage account information (e.g., for end user accounts) based on the rated financial information from the Rating Interface.

   ■ Billing/Payment Interface—deals with the transfer of funds.

■ Administrative Services—Although there are no specific requirements listed in OGSA Platform, it is expected that standard interfaces for performing administrative tasks (e.g., software upgrades, backups) in an automatic fashion will be desirable.

- Transactions—As grid solutions become deployed to support applications with requirements for the coordination of services and service state, there will be a requirement for a transaction service to do this coordination. This is more complicated in the grid environment than in the traditional distributed environment due to the potential latencies across geographies and the fluid nature of virtual organizations. The Web Services community is looking at this, and WS-Transactions has been proposed.
- Grid Service Orchestration—This refers to the coordination of a set of interacting services in pursuit of performing a single "task." This is also commonly known as workflow. The OGSA Platform is intending to define standard portTypes for launching workflows, as opposed to defining a "one size fits all" workflow language standard. Grid Service factories for different workflow systems can then be defined, allowing clients to use the most appropriate mechanism for their problem domain, yet access and control the workflow in a standard way.

9.4.4 OGSA Platform Models

The OGSI and OGSA Platform Interfaces provide basic mechanisms for managing Grid Services, and specifies some useful patterns for developing grid solutions, but on their own they do not actually provide any capability for dealing with the real entities (both virtual and physical), which are composed together to solve a particular computing problem. In order to bring it all together, standard representations of the entities on the grid need to be defined, so that the service data and operations in the implemented Grid Services actually have some real application. Thus, the OGSA Platform needs to include sets of common models for describing and manipulating the real entities that are represented through Grid Services. It is expected that there will be multiple models defined corresponding to specific domains. For example, a model for a "Linux HPC Cluster" might be defined, which describes the attributes of Linux clusters that are specific to HPC, and that are needed to run jobs on the cluster. A model for "Point of Sale" system might be defined, which includes services and attributes pertaining to this type of application (e.g., how to represent "orders" and "inventory" as Grid Services).

Currently, work is being done to define a model for representing manageable IT resources as Grid Services. Called the Common Resource Model (CRM),[10] it is intended to specify how to model the manageability of manageable resources. Manageable resources are defined as any entity in an IT system that has some form of state (run-time state, configuration, etc.). These entities can range from hardware,

---

[10] [CRM] *Common Resource Model (CRM)*, E. Stokes, N. Butler, Editors. Global Grid Forum. draft-ggf-crm-crmspec-1.

to software systems, to individual batch jobs. The CRM describes how management interfaces to these entities are represented as OGSA services. Each manageable entity is represented as a Grid Service, with state data being exposed through SDEs, and management operations represented through Grid Service operations.

CRM builds on existing standards for modeling resources, such as Distributed Management Task Force's (DMTF) Common Information Model (CIM) and various Management Information Base (MIB) specifications from IETF. The schemas defined here are then mapped to the WSDL and XSD schemas required to support the Grid Services model. CRM defines some additional XML Schema data types meaningful for management (e.g., counter and gauge), a base `portType` for manageable resources, and some of the mechanisms required to interact with these resources (e.g., aggregation through `ServiceGroup`, resource management lifecycle, etc).

9.5 BUILDING ON THE OGSA PLATFORM

Now that the notion of Grid Services are fairly well defined through the OGSA Platform, work needs to commence on defining the higher-level services which, when implemented, can actually be composed to do useful work. In order to identify some of these higher-level services, it is useful to examine the functionality of pre-OGSA grid deployments that are being used today. This includes the academic "*Partner Grids*" based on the Globus Toolkit, and production "*Enterprise Grids*" as implemented by Platform Computing within various industries including chip design, industrial manufacturing, the life sciences, and financial services. Because the many users of grid technologies today are still in the scientific and high performance computing fields, the use cases to draw on are generally based on some kind of computational analysis and simulation. Two essential elements of this kind of problem are the scheduling of computing resources (i.e., CPUs, memory, network, applications) and uniform access to data.

The Agreement-based Grid Service Management (WS-Agreement)[11] (OGSI-Agreement) specification defines the way that Grid Services can be managed based on organizational goals and application requirements, such that resources can be scheduled for use. The Data Access and Integration Services (DAIS)[12] working

---

[11] [WS-Agreement] *Agreement-based Grid Service Management (OGSI-Agreement), Version 0*, K. Czajkowski, A. Dan, J. Rofrano, S. Tuecke, Editors. Global Grid Forum. draft-ggf-czajkowski-agreement-00.

[12] [DAIS] *Services for Data Access and Data Processing on Grids*, V. Raman, I. Narang, C. Crone, L. Haas, S. Malaika, T. Mukai, D. Wolfson, C. Baru, Editors. Global Grid Forum.

group at the GGF is focused on developing specifications, which can help virtualize access to data sources, hiding the complexity associated with integrating data from multiple locations and in multiple forms. Most notable about both of these standards activities is that they are being equally driven from academia (where many of the issues have been defined and researched) and industry (where vendors of grid technologies are interested in deploying solutions that provide a value to a customer base). This greatly validates OGSA as a meaningful and useful specification. Vendors, such as Platform Computing (working on WS-Agreement), see that the adoption of standards are beneficial to their customers, and that OGSA specifically is the architecture of choice as a framework for future development.

9.5.1 WS-Agreement

Grid environments based on OGSA will be composed of many different, interacting Grid Services. Each of these services are in turn subject to different policies on how to manage underlying resources. In order to deal with the complexities of large collection of these services, there need to be mechanisms for Grid Service management.

From WS-Agreement, ". . . the ability to create Grid Services and adjust their policies and behaviors based on organizational goals and application requirements."

WS-Agreement defines the Agreement-based Grid Service Management model, which defines a set of OGSI-compliant `portTypes` allowing clients to negotiate with management services in order to manage Grid Services or other legacy applications (e.g., a local resource manager).

To put it in concrete terms, if a user wants to submit a compute job to run on a cluster, they would have their Grid Service client contact a job management service and would negotiate a set of agreements that ensured that the user's job would have access to a number of CPUs, memory, storage space, etc.

WS-Agreement defines fundamental mechanisms based on OGSI-compliant Agreement services, which represent an ongoing relationship between an agreement provider and an agreement initiator. The agreements define the behavior of a delivered service with respect to a service consumer. The Agreement will most likely be defined in sets of domain-specific agreement terms (defined in other specifications), as the WS-Agreement specification is focused on defining the abstraction of the agreement and the protocol for coming to agreement, rather than on defining sets of agreement terms.

Agreements are negotiated by creating a new Agreement service instance through a Grid Service that implements the `AgreementFactory` interface, inherited from Factory. A client calls the `Factory::createService` operation with `Creation-Parameters`, which represent the requested terms. Some `CreationParameters` can be required and some can be open to counter offers. If the agreement terms are not acceptable to the service provider, the `createService` operation returns a fault. Otherwise, a new Agreement service instance is created with SDEs representing the negotiated terms, or an `AgreementOffer` service instance is created, which represents a number of alternative offers that the client can then choose from. The instantiation of the Agreement service instance implies that the agreement has been accepted and is in force.

Referring back to the job submission example, the client might contact a job management service, which implements the `AgreementFactory` interface, with `CreationParameters` that say "my job has to have a software license for application X, would like to have 8 CPUs, and would like to have 4G of RAM." If the job management service could not provide the software license (the must have), the agreement terms would be rejected. If it could provide all of the terms, an Agreement service instance representing the job resources would be created. If, because of available resource constraints, the job management service could not fulfill the terms of the original `CreationParameters`, but could supply either 4 CPUs and 4G of RAM, or 8 CPUs and 2G of RAM, the job management service could create an `AgreementOffer` which included two potential agreements: "app X, 4 CPUs, 4G" and "app X, 8 CPUs, 2G," one of which the client could choose (because CPUs and memory were terms subject to counteroffers).

9.5.2 Data Access and Integration Services (DAIS)

The Data Access and Integration Services working group is focused on defining grid data services that provide consistent access to existing, autonomously managed databases. Although there had already been a lot of work around Grid Services for file management (e.g., GridFTP), database integration was not really covered by this work, even though databases play a central role in both the research and commercial computing domains.

When applications are ported to the grid, data access becomes even more of an issue. First, the wide area distribution of resources means that data to be processed could be located far away from the computation. Second, if a task running on the grid is accessing multiple data sources, with different formats and different access mechanisms, application developers spend a lot of time developing code to deal with these differences. In industry, the number one barrier to deploying grid solutions is often associated with the ability to access the right data sets in a consistent fashion across the grid.

The purpose of the DAIS specifications is to highlight a set of "transparencies," which are needed to deal with the complexities of data management, and then to define a set of OSGI-compliant Grid Services that is intended to virtualize data access, thus implementing the desired transparencies. The transparencies high-lighted are:

- *Heterogeneity* Transparency—applications accessing a data source should not have to be cognizant of the implementation of the data source. Common schemas help this, but there are often issues at the implementation level that get exposed to applications.
- *Name* Transparency—applications should not manipulate data objects directly, but access them through "logical domains" constrained by attributes. This implies both location and replication transparency.
- *Ownership and Costing* Transparency—applications should not need to separately negotiate access to data sources when dealing with multiple, autonomous data sources, both in terms of access rights and usage costs.
- *Parallelism* Transparency—applications processing data should automatically get parallel execution over nodes on the grid, if possible. Applications should be able to define workflows and interdependent tasks, and then have the grid fabric run the work in the most optimized way.
- *Distribution* Transparency—distributed data should be able to be maintained in a unified way.

DAIS then goes on to define the set of services needed in order to implement the transparency requirements. The following list shows how each service addresses the transparency requirements:

- Discovery—Name (maps logical domain and predicates onto actual data source or service, and chooses the best location and replica for a query), and Ownership and Costing (optimizes for source-independent metrics like time or cost).
- Federated Access—Heterogeneity (allows access independent of data format and implementation), Distribution (provides unified access to distributed data sources), and Name (provides some location transparency for file access).
- Consistency Management—Distribution (maintains consistency of distributed data).
- Collaboration—Distribution (maintains consistency of data updated in multiple places).
- Workflow Coordination— Parallelism (automatically parallelizes request).
- Authorization—Ownership and Costing (single sign-on).
- Replication/Caching—Name (migrates data on demand to the right place).
- Schema Management—Heterogeneity (maps between data formats).

9.6 IMPLEMENTING OGSA-BASED GRIDS

This section discusses some of the practical aspects of implementing OGSA-based grids.

9.6.1 The Globus Toolkit 3

The Globus Project is a consortium of academic researchers lead by principal investigators Ian Foster and Steve Tuecke from the Argonne (Illinois) National Laboratory together with Carl Kesselman from the Information Sciences Institute, University of Southern California (Marina del Rey). For about a decade, this group has been investigating distributed computing in support of Big Science—e.g., sharing *terascale* volumes of data from high-energy-physics (HEP) experiments between hundreds of globally distributed scientists, aggregating hundreds of CPUs to perform Grand Challenge computations, etc. In the latter half of the 1990s, their efforts focused on the Globus Toolkit—a software toolkit that addresses the key technical challenges in building grid tools, services, and applications.

The Globus Project adopted a pragmatic, bottom-up approach with the toolkit. The case of security provides an illustrative example. The Globus Project reviewed existing security solutions, and determined that none met all of their requirements. However, they determined that a Public Key Infrastructure (PKI) approach based around existing X.509 certificates, the Secure Sockets Layer[13] (SSL), certificate authorities, and a generic security API (GSS API), provided an excellent foundation. To this foundation, the Globus Project focused on developing extensions to support single sign on and delegation of rights—two of the essential requirements for the grid. The outcome was a viable security solution that made sensible use of known technologies—many of them standards in their own right.

Successful implementation and use of this security solution for the grid in the Globus Toolkit motivated activity in the standards arena. Through the Grid Security Infrastructure Working Group of the GGF, a specification for the Grid Security Infrastructure (GSI) was developed. This specification passed through the GGF's standardization process and gained approval. The case of security provides a representative example co-evolving specification and implementation. This co-evolution is a theme that persists today in the case of OGSA.

Figure 9.3 places GSI in the context of a protocol stack for version 2.x of the Globus Toolkit. It is evident that each of the higher-level components (e.g., GRAM, GridFTP, MDS) is based directly on a protocol (HTTP, FTP, LDAP). Although this also reflects the pragmatic, bottom-up approach taken at the outset, these underpinnings started to increasingly limit the ability to develop on top of the toolkit.

---

[13] SSL is now referred to as Transport Level Security (TLS).

With the development of the OGSI, which was driven greatly by the individuals involved in the Globus Project based on their experiences with Globus Toolkit 2.x, the next generation of the Globus Toolkit, referred to as version 3 (GT3), was intended to be the first reference implementation of the OGSI specification. Given the toolkit's open source license, GT3 would provide a basis for developers to start developing OGSA-based grid solutions.

On January 12th, 2003, at the GlobusWorld conference in San Diego, the Globus Project announced the availability of an alpha version of GT3. While preserving the functionality required for building grid solutions (i.e., GRAM, MDS, GridFTP), GT3 is based on a new infrastructure, which is compliant with OGSA and which implements the core functionality defined in OGSI.

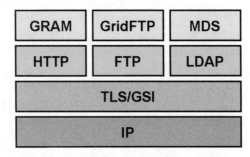

FIGURE 9.3   Protocol stack in version 2.x of the Globus Toolkit.

GT3 is composed of the following pieces:

- An OGSI reference implementation—Implements all the OGSI specified portTypes, as well as some APIs that help developers implement OGSI compliant services.
- Security Infrastructure—Based on GSI, this provides the encryption, authentication, and authorization services well known from GT2.
- System-level Services—Some infrastructural services that can be used with all other Grid Services. Services defined include Admin (for checking container availability and shutting it down), Logging (set log filters and aggregate log sources), and Management (for monitoring statistics regarding the health of the container environment).
- Base services—Higher level Grid Services, corresponding to GRAM, MDS, and GridFTP from the previous toolkit.

- User-defined services—Any higher-level services built on top of the base services. Perhaps contributed by non-Globus Project people.
- Grid Service Container—The OGSI runtime environment intended to shield end users from details of implementation, such as the kind of database used to store service data.
- Hosting environment—Implements the typical Web server type of services, such as transport protocols. Four hosting environments are supported: 1) an embedded environment for clients and lightweight servers, 2) a standalone server based on the embedded environment but with some additional server functionality, 3) the grid container inside a standard Java Servlet™ Engine, and 4) the grid container inside of an EJB application server.

GT3 Core focuses on providing support for writing Grid Services in the Java programming language. There are a number of utility classes intended to make it easier to develop OGSI compliant Grid Services. This includes support for automatically adding OGSI mandated service data to all services, APIs for dynamically adding service data definitions corresponding to SDEs in your WSDL, support for automatic lifecycle management for service instances in the container based on service demand, and some support for service state management.

With the completion of GT3, a complete OGSA environment can be depicted as in Figure 9.4.

9.6.2 GCSF

In the summer of 2003, Platform Computing contributed an open source implementation of the WS-Agreement protocol. Called the Grid Community Scheduling Framework (GCSF), it provides an implementation of the OGSI-Agreement Grid Services, built on the Globus Toolkit's OGSA hosting environment. The framework provides the basic protocols for negotiating agreements and the state data associated with managing agreements. The framework will not only provide the basis for the implementation of Grid Services for Meta Scheduling (i.e., the brokering of grid resources on behalf of a client) within Platform Computing's product suites, but for any developer who needs to define domain-specific agreement terms and needs an engine to manage them.

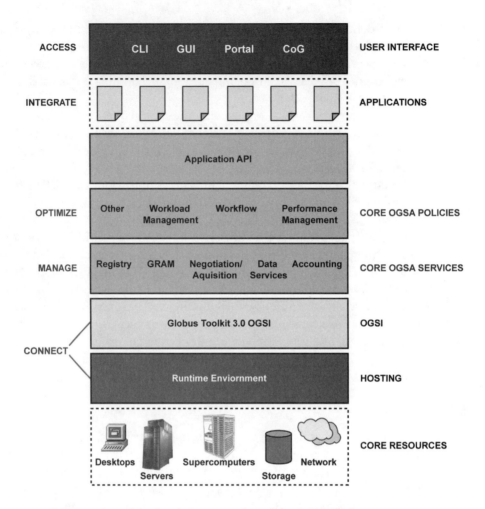

FIGURE 9.4 A functional representation of the OGSA Platform.

9.7 CONCLUSION

Much in the same way that COM changed the way office productivity solutions were implemented in the Microsoft Windows operating environment, OGSA will change the way that applications are developed to leverage grid environments. In summary, OGSA:

- Introduces an industry defined standard, encouraging modularity and facilitating the extensibility of an organization's grid environment.
- Facilitates the transition from large-scale, monolithic applications to collections of Grid Services that coordinate to provide application functionality.
- Enables users; clients do not need to be rewritten to take advantage of new and improved service implementations.
- Enables ISVs; leveraging existing OGSA defined interfaces simplifies and accelerates software development.

Open source implementations of OGSA such as the Globus Toolkit 3 make good on the promise of standards, by providing an implementation on top of which grid applications can be deployed. Furthermore, the promise of OGSA has been recognized within industry, demonstrated by specifications such as WS-Agreement, and the corresponding GCSF implementation from Platform Computing, showing that providers of commercial grid solutions see the OGSA standard as the vehicle to solve customer problems in the future.

10 Creating and Managing Grid Services

Omer F. Rana
Department of Computer Science and Welsh e-Science Centre, Cardiff University, UK

Ali Shaikhali
Department of Computer Science and Welsh e-Science Centre, Cardiff University, UK

Gregor von Laszewski
Mathematics and Computer Science Division, Argonne National Laboratory, U.S.

In This Chapter

- Creating Grid Services
- Managing Grid Services
- Discovering Grid Services

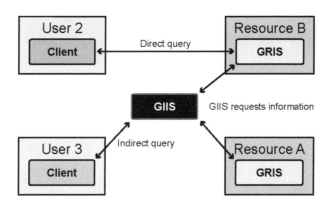

10.1   INTRODUCTION

Many definitions exist for what constitutes a "grid"—the most common of which identifies a *Computational Grid* as a collection of distributed computing resources available over local or wide area networks which appear to an end user as one large virtual computing system. A significant effort has been invested by the grid community, academic as well as commercial, into building software systems that may be used to manage such a virtual computing system. The grid approach is an important development for both the business and the scientific community[1] and promotes a vision for sophisticated scientific and business collaborations. Until recently, most of this effort was focused on managing resource ensembles and coordinating access to such ensembles in a secure and efficient manner. The focus of the technology discussed in this chapter was driven primarily by the need to handle differences among different kinds of hardware architecture (Intel vs. Sun vs. SGI), operating systems (Solaris™ vs. Win2000™ vs. IRIX™), and data repositories (structured databases vs. file systems). To establish a "virtual organization," it is also necessary to allow individual systems administrators managing one or more resource ensembles to exercise their individual authority over their resources without allowing some centralized system to override access rights that they have defined for external users coming to their system. The predominant interest within existing grid systems remains on "resources," the hosting platforms, and storage devices that may be combined to establish a virtual system. Increasingly however, there has been a recognition that it is also important to capture details about software programs and libraries that are run on such resources.

10.2   SERVICES AND THE GRID

Consequently, the grid community has recently expressed interest in "service-oriented" computing. Services are seen as a natural progression from component-based software development[2] and as a means to integrate different component development frameworks. A service in this context may be defined as a behavior that is provided by a component for use by any other component based on a network-addressable interface contract (generally identifying some capability provided

---

[1] [Laszewski03] G. von Laszewski, G. Pieper, and P. Wagstrom, "Gestalt of the Grid," in *Performance Evaluation and Characterization of Parallel and Distributed Computing Tools*, Wiley Book Series on Parallel and Distributed Computing, to be published 2003. [Online]. Available: *http://www.mcs.anl.gov/~laszewsk/bib/papers/vonLaszewski—gestalt.pdf*

[2] [Stevens02] Service-Oriented Architecture Introduction, Part 1. See Web site at *http://software-dev.earthweb.com/Microsoft/article/0,,10720_1010451_1,00.html*

by the service). In the simplest case, such a contract specifies the set of operations that one service can invoke on another (a set of method calls, for instance, when a service is implemented by using object-oriented technologies). Generally, such an interface may also contain additional attributes that are not specifically related to the functionality of the called service and can include aspects such as the performance of the service, the cost of accessing and using the service, and the details of ownership and access rights associated with a service. The interface also allows the discovery, advertising, delegation, and composition of services. Consider an automobile company that has the need to share its compute and data resources among its employees to enable them to share and access these resources as part of an international collaborative design process for developing a new car. We can identify the following participants involved in such a collaboration:

- Grid services and toolkit developers—These users care about what grid tools and resources are available within the grid infrastructure to guide the software development process.
- Grid administrators—These users want to know about the status of resources in a production mode setting, which includes controlling, monitoring, and utilization-related information.
- Grid application users—These users care about having a transparent high-level view of the grid that exposes information related to a convenient problem-solving environment rather than the maintenance and development of complex grid applications.

Hence, a grid information framework not only needs to support a diverse set of information but also needs to serve a diverse set of grid users.

A service stresses interoperability and may be dynamically discovered and used. According to the OGSA specification,[3] the service abstraction may be used to specify access to computational resources, storage resources, and networks in a unified way. How the actual service is implemented is hidden from the user through the service interface. Hence, a compute service may be implemented on a single- or multi-processor machine, but these details may not be directly exposed in the service contract. The granularity of a service can vary; a service can be hosted on a single machine, or it may be distributed. The TeraGrid project[4] provides an example of the use of services for managing access to computational and data resources. In this project, a computational cluster of IA-64 machines may be viewed as a compute service, for instance, thereby hiding details of the underlying operating system

---

[3] [Physiology] The Physiology of the Grid: An Open Grid Services Architecture for Distributed Systems Integration, I. Foster, C. Kesselman, J. Nick, S. Tuecke

[4] [TERA] TeraGrid, *www.teragrid.org*

and network. A developer would interact with such a system using the Open Grid Services Architecture (OGSA) toolkit, derived from the Globus Toolkit,[5] and consisting of a collection of services and software libraries. Hence, it is assumed that a computational grid is composed of a number of heterogeneous resources which may be owned and managed by different administrators. Each of these resources may offer one or more services:

- A single application with a well-defined API (which also includes wrapped applications from legacy codes). In this case, the software application is offering some service (which may vary in granularity from a *simulation kernel* to a *molecular dynamics application*, for instance). In order to run such an application, it may be necessary to configure the operating environment (such as set up PATH variables that point to the location of particular libraries), and make available third-party libraries.
- A single application used to access services on other resources managed by a different systems administrator.
- A collection of coupled applications, with predefined interdependencies between elements of the collection. Each element provides a subservice that must be combined with other services in a particular order.
- A software library containing a number of subservices all of which are related in some functional sense—e.g., a graphics or a numerics library.
- An interface for managing access to a resource. (This may include access rights and security privileges, scheduling priorities, and license checking software.)

A distinguishing feature of a service is that it does not involve persistent software applications running on a particular server. A service primarily is executed only when a request for the service is received by a *service provider*. The service provider publishes (advertises) the capability it can offer, but it does not need to have a permanently running server to support the service. The environment that supports the service abstraction must satisfy all preconditions needed to execute a service. A service may also have *soft state*, implying that the results of a service may not exist forever. Soft state is particularly important in the context of dynamic systems such as Computational Grids, where resources and user properties can vary over time. The soft state mechanism also allows a system to adapt its structure, depending on the behavior of participants as they enter and leave a system at will. The soft-state approach allows participation for a particular time period, subsequently requiring the participants to renew their membership. Participants not able to renew their membership are automatically removed from the system.

---

[5] [Globus] The Globus Project, *www.globus.org*

OGSA (see Chapter 9) is a distributed interaction and computing architecture that aims to integrate grid systems with Web Services, and it defines standard mechanisms to create, name, and discover Grid Service instances. The core concept within this architecture is the notion of a "Grid Service," defined in the Web Services Description Language (WSDL), and containing some pre-defined attributes (such as types, operations, and bindings). WSDL provides a set of well-defined interfaces that require the developer to follow specific naming conventions. A new tag `gsdl` has been added to the WSDL document to enable description of Grid Services. Once the interface of a Grid Service has been defined, it must be made available for use by others. This process generally involves publishing the service within one or more *registries*. The standard interface for a Grid Service includes multiple bindings and implementations (such as the Java and C# languages), and development may be undertaken by using a range of commercial (such as Microsoft's Visual Studio.NET™) or public-domain (IBM's WSTK) tools. A Grid Service may be deployed on a number of different hosting environments, although all services require the existence of the Globus Toolkit. Grid Services implemented with OGSA are generally transient and are created by using a factory service. Hence, there may be many instances of a particular Grid Service, and each instance can maintain internal state. Service instances can exchange state via messaging. Figure 10.1 indicates how a user may access Grid Services via a portal interface. Connectivity between the user and a number of organizations is achieved by using XML/SOAP messages. Each organization in this instance offers a particular set of services—such as mathematical and graphics libraries—that are encoded as WSDL services. A user can pick and combine a number of such services to implement an application. Each organization may itself support a local grid containing compute and data resources internal to an organization. The ability to integrate a number of local grids is a significant advantage of using Web Service technologies to build Computational Grids.

When creating a new service, a user application issues a `create Grid service` request on a factory interface, leading to the creation of a new *instance*. This newly created instance will now be allocated some computing resources automatically. An initial lifetime of the instance can also be specified prior to its creation and allows the OGSA infrastructure to keep the service "alive" for the duration of this instance. This is achieved by sending it `keepalive` messages. The newly created instance is assigned a globally unique identifier called the Grid Service Handle (GSH)—and is used to distinguish this particular service instance. The implementation of a Grid Service involves the following steps:[6]

---

[6] [OGSADF] Thomas Sandholm and Jarek Gawor, "Grid Services Development Framework Design," part of the *OGSA pre-release*. Available at: *http://www.globus.org/ogsa/releases/TechPreview/ogsadf.pdf*

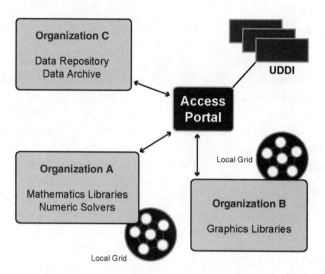

FIGURE 10.1 General framework for enabling interaction between multiple grid systems based on Web Services.

- Write a WSDL PortType definition, using OGSA types (or defining new ones).
- Write a WSDL binding definition, identifying ways in which one could connect to the service, e.g., by using SOAP/HTTP, TCP/IP, etc.
- Write a WSDL service definition based on the PortTypes supported by the service and identified in Step 1.
- Implement a factory by extending the FactorySkeleton provided, to indicate how new instances of a service are to be created.
- Configure the factory with various options available, such as schemas supported.
- Implement the functionality of the service by extending the ServiceSkeleton class. If an existing code (legacy code) is to be used in some way, then the *delegation* mechanism should be used. When used in this mode, the factory returns a skeleton instance in Step 4.
- Implement code for the client that must interact with the service.

Extensibility elements available within WSDL play a significant role in OGSA, enabling users to customize their Grid Services. This capability is useful to enable a community or group of users to define special XML tags or queries that may be relevant within a particular application context. In OGSA, such extensibility elements include *factory input parameters*, *query expressions*, and *service data elements (SDEs)*. Factory input parameters provide the user with a mechanism to customize the creation of a new Grid Service. Query expressions enable a user to utilize a specialist query language or representation scheme. SDEs enable the definition of specialist XML tags. A mathematical service which makes use of numeric service bindings is illustrated in code Listing 10.1.

Listing 10.1

```xml
<?xml version="1.0" encoding="UTF-8"?>
<definitions name="MathServiceDefinition"
 targetNamespace="http://samples.ogsa.globus.org/math"
xmlns="http://schemas.xmlsoap.org/wsdl/"
xmlns:grid-service-
bindings="http://ogsa.gridforum.org/service/grid_service_bindings"
xmlns:soap="http://schemas.xmlsoap.org/wsdl/soap/">

<import location="numeric_bindings.wsdl"
 namespace="http://samples.ogsa.globus.org/math/numeric_bindings"/>

<import location="../service/grid_service_bindings.wsdl"
namespace="http://ogsa.gridforum.org/service/grid_service_bindings"/>

<service name="MathMatrix">
  <documentation> ... </documentation>

<port binding="numeric_bindings:NumericSOAPBinding"
 name="NumericPort">
 <soap:address location="http://localhost:8080/ogsa/services"/>
</port>

<port binding="grid-service-bindings:GridServiceSOAPBinding"
 name="GridServicePort">
 <soap:address location="http://localhost:8080/ogsa/services">
</port>
</service>
</definitions>
```

10.3 CONVERTING EXISTING SOFTWARE

A number of tools have emerged recently that enable automatic creation of Web Service interfaces (WSDL- and SOAP-based) from existing Java-based implementations, such as SOAPswitch[7] and Java2WSDL in the Web Service Toolkit.[8] Such tools are essential to bootstrap interest in Web Services by allowing existing applications to be made available as Web Services. Generally, there is little interest in rewriting an existing application as a Web or Grid Service because significant investment has already been made in designing, implementing, and maintaining existing software. This situation is particularly relevant for the product development, science, and engineering community, where a numeric solver or a visualization routine, for instance, has been developed over many years, and conversion to a new technology often requires a considerable investment. Further, many designers of the initial system may not be available, so the exact reasons for particular design choices are not apparent (and often not well documented). The uptake of Grid Services within such a community is therefore also likely to be constrained by how existing codes may be efficiently reused, and suitable tools are needed to prevent rewrites of existing codes. Wrapping of existing codes as services may be undertaken at different levels:

- Wrapping executables—This is the wrapping "as-is" approach, and is important when no source code is available. The execution environment of the original must be maintained, such as any particular dynamic link libraries that need to be referenced during execution.
- Wrapping source—This is the "source-update" approach, whereby additional code needs to be written to interface with the existing application (generally the I/O routines). The original executable program needs to relinquish some control to the wrapper. This approach also requires some data type conversions to be undertaken between the original and the new types supported by the wrapper. Web Services make this process somewhat simpler, as a developer can specify a namespace to declare particular data types present in the original code and can subsequently reference these in the wrapper.
- Source-split wrapping—This is the "unit-wrapping" approach, where the source code is divided into individual units and can be wrapped separately. This approach is particularly relevant when the code being wrapped is a software library or contains a collection of independent routines.

---

[7] [actional02] Actional—Web Services Management Solutions. "SOAPswitch." See Web site at *http://www.actional.com/products/soapswitch/index.asp*. Last visited: October 2002.

[8] [wstk] IBM—Alphaworks, "Web Services Toolkit." See Web site at *http://www.alphaworks.ibm.com/tech/webservicetoolkit/*. Last visited October 2002.

■ Application-supported wrapping—This is the "app-wrap" approach, whereby the wrapper is adding new functionality to the code and not simply acting as a mediator between the Web Service and the existing application.

Regardless of which approach is adopted, the aim is to generate a WSDL document for every identified service, the PortType, and the associated bindings. Additional properties about the service may be encoded in the service data elements. An alternative approach, adopted in the Web Services Invocation Framework,[9] is to provide multiprotocol bindings enabling automatic invocation of the relevant binding based on the content carried in the request. Hence, bindings are providing for COM objects, HTTP requests, and so forth. Once a WSDL interface has been defined, it is important to place this into a registry service, so that other applications can automatically discover and make use of the service. In the context of Web Services, the discovery mechanism is generally provided by the Universal Description, Discovery and Integration (UDDI) registry, which may be organized in the same way as domain name servers. Hence, services may register with one or more registry services (using the same identifier), and may be discovered by search distributed over one or more registries.

Current UDDI implementations are limited in scope, in that they allow search to be carried out only on limited attributes of a services, namely, on service name (which may be selected by the service provider), or key Reference (which must be unique for a service), or based on a categoryBag (which lists all the business categories within which a service is listed.[10] Furthermore, public UDDI registries may contain listings for business that no longer exist or sites that are no longer active.[11] The interaction between UDDI registries is also an importance concern, and there is no main consensus at the present time of who should own the root UDDI registries. Extensions to UDDI, such as "UDDIe," address some of these restrictions: (1) support for "leasing," to enable services to register with UDDI for a limited time period (to overcome the problem of missing or inconsistent links), (2) support for search on other attributes of a service, achieved by extending the businessService class in UDDI with propertyBag, and (3) extending the find method to enable queries to UDDI to be based on numeric and logical (AND/OR) ranges. The extensions allow UDDIe to co-exist with existing UDDI implementations and enable a query to be issued to both simultaneously. In subsequent sections, we describe these extensions and their use.

---

[9] [wsif] IBM—Alphaworks, "Web Services Invocation Framework." See Web site at: *http://www.alphaworks.ibm.com/tech/wsif/* . Last visited October 2002.

[10] [uddiwhitepaper] UDDI Technical White Paper. Available at: *http://www.uddi.org/pubs/Iru_UDDI_Technical_White_Paper.pdf*

[11] [Mello02] Adrian Mello, "Breathing New Life into UDDI," Tech Update, ZDNET.com, June 26, 2002.

10.4 SERVICE DISCOVERY

Service discovery involves search through local and remote registries, to discover services that match a particular criterion defined by the user. In Grid Computing, this was generally restricted to hardware resources being made available as services and registered in a directory service such as the Metacomputing Directory Service (MDS) (now renamed the Monitoring and Discovery Service). The MDS uses the Lightweight Directory Access Protocol (LDAP) to hierarchically structure the available resources within organizational domains. Each node within the LDAP structure contains properties associated with a particular resource. A user or application program can query the LDAP server to discover the type of hardware architecture (an Intel or a Sun machine, for instance) the operating system supported and details about the job manager that is available at a given resource. A variety of query terms are available as grid_info_search, which can be configured to discover a particular property of a resource. Each organizational domain participating in the grid would need to run a Grid Resource Information Service (GRIS) that is registered with a Grid Index Information Service (GIIS) to form a virtual information aggregation. The GIIS/GRIS together constitutes the discovery service currently provided in the Globus Toolkit. Hence, service discovery is supported via an "Information Service" that must relay information about grid infrastructure to its users. All of these user communities require an elementary set of functionality to be supported by the grid information service. This functionality includes

- Lookup and retrieval of information that can be pinpointed to a particular resource or object
- Query and search of information to retrieve a collection of related resources or objects
- Creation of new information upon request that is otherwise not stored or cached in the grid
- Event forwarding to relay the information based on dynamic events within the grid infrastructure
- Aggregation for topical retrieval and organization of information
- Filtering of information to reduce the amount or useful information communicated between services, to increase performance and reduce the amount of information to be stored
- Storage, backup, and caching of the information to make them available for easier retrieval or enhance the performance
- Security, protection, and encryption in order to enable important security related activities such as access control, authentication

In order to support these functionalities, there are several services that may be used by the grid users. Figure 10.2 lists a number of corresponding services implementing such functionality. We note that there may be many different services based on the diverse user communities and needs in each category. Usually, some high-level services are provided that combine certain aspects of the base functionality. A common example is the implementation and use of a directory service to support the functionalities of lookup and retrieval of information that is classified in a hierarchy. Such a service needs to provide support for both white and yellow pages.

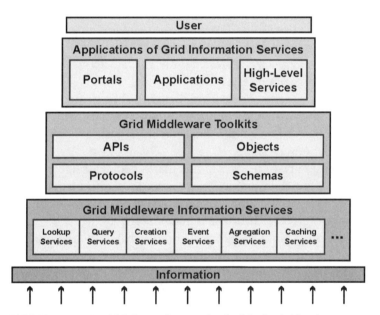

FIGURE 10.2 A grid information service builds the bridge between information about resources on the grid and the user community.

10.5 OPERATIONAL REQUIREMENTS

To choose an appropriate implementation strategy for a grid information service, one must consider a number of important requirements. Some of these requirements such as scalability and performance can be expressed in quantitative terms; others such as expressiveness, deployability, and performance are more subjective. Typically, use patterns for grids operate on a large scale (hundreds and thousands of resources) and have demanding performance requirements. Hence, an information infrastructure must permit rapid access to frequently used and dynamically changing information. We list such operational requirements that include capability, security, and scalability concerns as identified in Laszewski:[12]

■ Scalability and cost—The infrastructure must scale to large numbers of components and permit concurrent access by many entities. At the same time, its organization must permit easy discovery of information. The human and resource costs (CPU cycles, disk space, network bandwidth) of creating and maintaining information must also be low, both at individual sites and in total.

■ Uniformity—The goal should be to simplify the development of tools and applications that use data to guide configuration decisions. A uniform data model is required, as well as an application programming interface (API) for common operations on the data represented via that model. One aspect of this uniformity is a standard representation for data about common resources, such as processors and networks.

■ Expressiveness—The data model should be rich enough to represent relevant structures within distributed computing systems. A particular challenge is representing characteristics that span organizations—for example, network bandwidth between sites.

■ Extensibility—Any data model that is defined will, by necessity, be incomplete. Hence, the ability to incorporate additional information is important. For example, an application can use this facility to record specific information about its behavior (observed bandwidth, memory requirements) for use in subsequent runs.

■ Diversity (Multiple information sources)—The required information may be generated by many different sources. Consequently, an information infrastructure must integrate information from multiple sources.

■ Dynamicity—Some of the data required by applications are highly dynamic—for example, network availability or load. An information infrastructure must be able to make these data available in a timely fashion.

■ Flexibility—It is necessary to both read and update data contained within the information infrastructure. Some form of search capability is also required, to assist in locating stored data.

■ Security—It is important to control who is allowed to update configuration data. Some sites will also want to control access.

■ Deployability—An information infrastructure is useful only if is broadly deployed. In the current case, we require techniques that can be installed and maintained easily at many sites.

■ Decentralized maintenance—It must be possible to delegate the task of creating and maintaining information about resources to the sites at which re-

---

[12] [Laszewski97] S. Fitzgerald, G. von Laszewski, I. Foster, C. Kesselman, W. Smith, and S. Tuecke, "A Directory Service for Configuring High-Performance Distributed Computations," in *Proceedings of the 6th IEEE Symposium on High-Performance Distributed Computing*, 5-8 Aug. 1997, pp. 365–375. [Online]. Available: *http://www.mcs.anl.gov/~laszewsk/papers/fitzgerald—hpdc97.pdf*

sources are located. This delegation is important for both scalability and security reasons.

10.6 TOOLS AND TOOLKITS

Currently, the Globus Toolkit 3 provides a de facto standard toolkit for developing prototype Grid Services applications. Besides including an information service, the Globus Toolkit includes services for remote job submission, file transfer, and security. Because the Globus Toolkit supports a set of protocols, one can interface with these services through a variety of client libraries. Easy to use client libraries are provided in Java and Python as part of the Commodity Grid (CoG) Kit project.

10.6.1 Globus Toolkit Grid Information Service

As mentioned earlier, the Globus Toolkit contains a grid information service called MDS. Initially an acronym for Metacomputing Directory Service, MDS now denotes Monitoring and Directory Service to better reflect that MDS is more than an information service for metacomputers.[12, 13, 14, 15]

MDS was designed from the start as a distributed information service with an information entry point for each virtual organization. This model has been further refined over the years and provides by now also individual information services running on the resources. The information model used in the MDS is based on entries arranged in a directory information tree with object entries. Object classes describe what information can be stored in the directory. For a detailed discussion about object classes, entries, and their attributes, refer to the MDS schema Web page.[16]

The MDS comprises the Grid Resource Information Service (GRIS) and Grid Index Information Service (GIIS) as depicted in Figure 10.4. A GRIS is an information service that runs on a single resource and can answer queries from a user

[13] [MDSuser] "MDS 2.2 User's Guide," *http://www.globus.org/mds/mdsusersguide.pdf*

[14] [Laszewski02] G. von Laszewski, J. Gawor, C. J. Pena , and I. Foster, "InfoGram: A Peer-to-Peer Information and Job Submission Service," in Proceedings of the 11th Symposium on High Performance Distributed Computing, Edinburgh, U.K., 24-26 July 2002, pp. 333–342. [Online]. Available: *http://www.mcs.anl.gov/laszewsk/papers/infogram.ps*

[15] [Czajkowski01] K. Czajkowski, S. Fitzgerald, I. Foster, and C. Kesselman, "Grid Information Services for Distributed Resource Sharing," in 10th IEEE International Symposium on High Performance Distributed Computing. San Fransisco, CA: IEEE Press, August 7-9 2001, pp. 181–184, *www.globus.org*

[16] [MDSschema] "MDS Schema," Web page. [Online]. Available: *http://www.globus.org/mds/ Schema.html*

about that particular resource by directing these queries to an information provider[17] deployed on that resource. An information provider is a service that generates information about a specific aspect of a resource. The query from GRIS to a resource typically contains such information related to the resource platform, architecture, operating system, CPU, memory, network, and file systems. One can, however, build customized information providers that return additional information. GIIS is an aggregate directory service that builds a collection of information services out of multiple GIIS. It supports queries against information spread across multiple GRIS resources.

Hence, GRIS can respond to queries from other systems on the grid asking for information about a local machine or other specific resource. It can be configured to register itself with aggregate directory services such as GIIS (so that those services can pass on information about the machine to others). GRIS au-

---

[17] [MDSGRIS] "GRIS Information Providers," Web page. [Online]. Available: *http://www.globus.org /mds/DefaultGRISProviders.html*

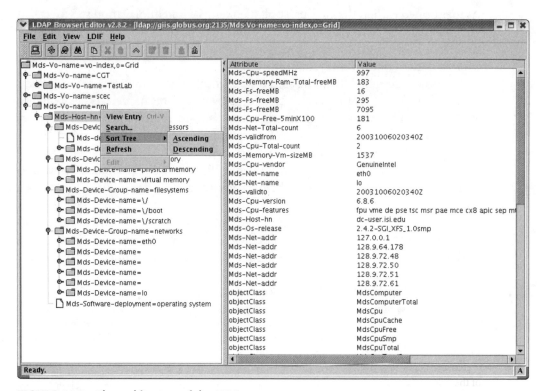

FIGURE 10.3 The architecture of the MDS.

thenticates and parses each incoming information request and then dispatches those requests to one or more local information providers, depending on the type of information named in the request. Results are then sent back to the client. In order to get collective information about two or more resources present in a single site, the queries can be sent directly to GIIS. In that case, the GIIS directs the query to GRIS.

MDS can use the Grid Security Infrastructure (GSI), which enables the use of certificates for authentication and authorization. MDS provides both authenticated and anonymous access by users. Site policies specify the restrictions on registration of resources with GIIS by the system administrator. An open policy for a GIIS allows all a GRIS or GIIS resources to be viewed by the users, whereas in a closed system only specified resources can be viewed and can register with a GIIS.

10.6.2 Accessing Grid Information

This section explains the different methods provided by the Java CoG Kit[18, 19] to access the grid information services. It will briefly introduce tools for accessing the information through an information browser, portals, and Java API, and shell scripts.

The Java CoG Kit project developed a LDAP browser/editor that provides a user-friendly Windows Explorer™-like interface to LDAP directories with tightly integrated browsing and editing capabilities. It is written entirely in Java with the help JNDI class libraries. It can connect to LDAP v2 and v3 servers. Figure 10.4 shows the user interface of the LDAP browser/editor. Other interfaces have been developed in Perl, PHP, and Python. However the LDAP browser/editor is currently the most advanced interface.

In addition to this high-level graphical user interface, the Java CoG Kit provides a shell or batch script, called grid-info-search, that is available for the UNIX and Microsoft Windows operating systems. It allows elementary lookup and search operations similar to standard LDAP queries. By providing the proper parameters, one can list all objects from a GRIS service. For example, the following command can be used to retrieve all available information from the machine hot.mcs.anl.gov while accessing a GRIS service running at port 2135:

---

[18] [Laszewski01] G. von Laszewski, I. Foster, J. Gawor, and P. Lane, "A Java Commodity Grid Kit," *Concurrency and Computation: Practice and Experience*, vol. 13, no. 8–9, pp. 643–662, 2001. [Online]. Available: *http://www.globus.org/cog/documentation/papers/cog-cpe-final.pdf*

[19] [Laszewski02a] G. von Laszewski, J. Gawor, P. Lane, N. Rehn, M. Russell, and K. Jackson, "Features of the Java Commodity Grid Kit," *Concurrency and Computation: Practice and Experience*, vol. 14, pp. 1045–1055, 2002.

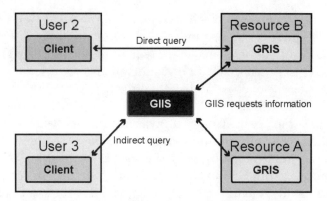

FIGURE 10.4   The Java CoG Kit LDAP browser/editor.

```
> grid-info-search -x -h hot.mcs.anl.gov -p 2135
                          -b "Mds-Vo-name=local, o=Grid"
  "(objectclass=*)"
```

The option -x is used to denote anonymous access, and -b specifies the location in the directory from which to start the search. The final argument in this example is the search filter; it specifies the category of object class one wishes to search. A partial output from this command looks as follows:

```
dn: Mds-Host-hn=hot.mcs.anl.gov, Mds-Vo-name=local, o=Grid
Mds-Cpu-speedMHz: 866
Mds-Memory-Ram-Total-freeMB: 304
Mds-Fs-freeMB: 4428
Mds-Cpu-Free-5minX100: 134
Mds-Net-Total-count: 2
Mds-validfrom: 20030303165825Z
Mds-Cpu-Total-count: 2
Mds-Memory-Vm-sizeMB: 243
Mds-Cpu-vendor: GenuineIntel
Mds-Net-name: eth0
Mds-Net-name: lo
Mds-validto: 20030303165825Z
...
```

Specific attribute value pairs can be selected by appending the appropriate attribute names to the search command, as indicated in the next example.

```
> grid-info-search -x -h giis.mcs.anl.gov -p 2135
         -b "Mds-Vo-name=site, o=Grid"
         "(&(objectclass=MdsCpu)(Mds-Host-hn=cold.mcs.anl.gov))"
         Mds-Cpu-model Mds-Cpu-speedMHz
```

Because the MDS uses the LDAP protocol, one can easily develop programs in Java through Java Naming and Directory Interface (JNDI). JNDI is a generic API distributed with J2EE for retrieving directory information.

A simple example of using JNDI for accessing the information in MDS is provided next. It requires only a couple of elementary operations that start with establishing the connection and end in retrieving and displaying the information.

```
// Step 1: Set MDS host and port information
Hashtable env = new Hashtable();
env.put(Context.PROVIDER_URL, "ldap://"+ host + ":" + port);
// Step 2: Specify the anonymous access
env.put(Context.SECURITY_AUTHENTICATION, "simple");
// Step 3: Create the Initial Dir Context
DirContext ctx = new InitialDirContext(env);
// Step 4: Search for required information
String baseDN = "mds-vo-name=local, o=grid";
String filter = "(objectclass=*)";
NamingEnumeration results = ctx.search(baseDN, filter, null);
// Step 5: Display the results
SearchResult result; Attributes attributes;
while (results.hasMoreElements()) {
    result = (SearchResult)results.next();
    attributes = si.getAttributes();
    System.out.println(si.getName() + ":");
    System.out.println(attrs);
    System.out.println();
}
```

The Java CoG Kit also provides a security mechanism that is portable with the Globus Toolkit 2. Hence, one can design and enable an access control list-enabled GIS service. It is important to note that this security library is currently used also in OGSA making the Java CoG Kit an integral part of future Globus Toolkit releases. For details, see the Java CoG Manual.[20]

---

[20] [CoGmanual] G. von Laszewski, B. Alunkal, K. Amin, J. Gawor, M. Hategan, and S. Nijsure, "The Java CoG Kit User Manual," Argonne National Laboratory, Mathematics and Computer Science Division, 9700 S. Cass Ave, Argonne, IL 60439, U.S.A., MCS Technical Memorandum ANL/MCS-TM-259, Mar. 14, 2003.

10.6.3 Performance Issues with MDS

The performance of a query depends upon the information providers and the frequency of retrieval. If the data is cached, the data will be returned very quickly. In case the data is stale, a new request for obtaining the data is issued and the query returns only after the value is updated in the cache.

Retrieving information from MDS should be performed with thought and care. One should connect to the MDS server only as long as the connection is required, in order to avoid blocking the limited number of ports to an MDS server. Because a connection usually takes time, it is sometimes better to perform a number of queries. One should, however, avoid analyzing results between subsequent queries. Instead one should analyze the queries once all queries have been performed, or one should start parallel query threads.

Additionally, a user needs to think about the correlation between query frequency and information update frequency of a value in the MDS. For example, if a user requests information every second that is updated only every 30 seconds, resources will be wasted. Grid programmers can avoid such situations by having a clear understanding of how the information is updated.

10.6.4 Other Information Services and Providers

Grid projects that use information services include Condor,[21] Legion,[22] and Netsolve.[23] Tools such as the Network Weather Service (NWS),[24] Ganglia,[25] and even the Unix command sysinfo can be integrated into a Globus-based grid infrastructure by using appropiate Globus Toolkit information providers.

---

[21] [CONDOR] M. J. Litzkow, M. Livny, and M. W. Mutka, "Condor—A Hunter of Idle Workstations," in Proceedings of the 8th International Conference on Distributed Computing Systems (ICDCS). San Jose, California: IEEE Computer Society, June 1988, pp. 104–111. [Online]. Available: *http://www.cs.wisc.edu/condor/*

[22] [LEGION] A. S. Grimshaw, W. A. Wulf, and the Legion team, "The Legion Vision of a Worldwide Virtual Computer," Communications of the ACM, vol. 40, no. 1, pp. 39–45, 1997.

[23] [NETSOLVE] H. Casanova and J. Dongarra, "NetSolve: A Network Server for Solving Computational Science Problems," *International Journal of Supercomputer Applications and High Performance Computing*, vol. 11, no. 3, pp. 212–223, 1997.

[24] [NWS] B. Gaidioz, R. Wolski, and B. Tourancheau, "Synchronizing Network Probes to avoid Measurement Intrusiveness with the Network Weather Service," in Proceedings of 9th IEEE High-Performance Distributed Computing Conference, August 2000, pp. 147–154.[Online]. Available: *http://www.cs.ucsb.edu/rich/publications/*

[25] [GANGLIA] M. L. Massie, B. N. Chun, and D. E. Culler, "The Ganglia Distributed Monitoring System: Design, Implementation, and Experience," *Parallel Computing*, 2003, (submitted). [Online]. Available: *http://ganglia.sourceforge.net/talks/parallel computing/ganglia-twocol.pdf*

10.6.5 Future

An early glimpse into the future of grid information services was published in 2001.[26] Here, a radical new approach was presented resulting in a framework that showed how to serve job submissions and information queries from the same service while using a common specification language. Additionally, several other features such as advanced caching and the introduction of a quality-of-information attribute have been proposed as part of the functionality of a grid information service.

The Java CoG Kit has significantly helped in the development of OGSA, as initial versions of OGSA are clearly based on the security, file transfer, and job execution services distributed with the Java CoG Kit. Currently, efforts are under way to distribute the Java CoG Kit also as part of the Globus Toolkit version 3 (see Chapter 9).

10.7 SUPPORT IN UDDI

An alternative approach to discovery is provided in the UDDI registry, where UDDI provides a standard method for publishing and discovering information about Web Services. The UDDI Project is an industry initiative that attempts to create a platform-independent, open framework for describing and discovering services offered by one or more businesses. UDDI is therefore more generic in nature—compared with GRIS/GIIS—and is also relevant in the context of storing details of Grid Services. The particular issues that UDDI attempts to address include:

■ Making it possible for organizations to quickly discover the right business from millions of others currently online
■ Defining how to enable interaction with the business to be conducted, once it has been discovered

These issues are intended to increase the ability of a business to discover new customers and expand its market reach. Service discovery can happen at application design time or at run time. For instance, a user may search a service registry via a browser interface and manually evaluate the results. Alternatively, the application may itself issue a `find` request to the registry and use analysis logic to evaluate the results. Current efforts in UDDI development, such as work being undertaken by

---

[26] [Laszewski01] G. von Laszewski, I. Foster, J. Gawor, and P. Lane, "A Java Commodity Grid Kit," *Concurrency and Computation: Practice and Experience*, vol. 13, no. 8–9, pp. 643–662, 2001. [Online]. Available: *http://www.globus.org/cog/documentation/papers/cog-cpe-final.pdf*

the OASIS technical committee[27] address issues of security and access privileges, ranging from mechanisms for encrypting SOAP messages to support for a security policy that mediates interaction between Web Services. Work is also underway to add features to UDDI, such as providing support for new match mechanisms (such as `caseInsensitiveMatch`, `diacriticSensitiveMatch`) and the capability to arrange services in UDDI based on various sort criteria.

UDDI is a registry and *not* a data repository and therefore should not be used to store data values related to a particular service but rather the metadata about the structure of the data. In the context of Grid Computing, a registry may contain the WSDL documents about particular services, the current location of services, a fault or audit log identifying error conditions that occurred on particular services, or access rights about one or more services, for instance. It should not be used to record input or output data generated from a service. The primary reason is to keep registry services lightweight and enable ease of administration and replication of content within them. However, a registry may have references to real data stored in a structured database or a file system. There may be two types of UDDI registries: those that are private to a given organization and those that are to be made available for public use. Private registries may contain services that are useful only in a local context and that may be hidden behind company firewalls. An organization may also utilize private registries to record temporary information about processes active at a given time. Public registries contain services that may be discoverable by other users and that must be advertised to external users. UDDI registries may also be arranged in a hierarchy, whereby public registries available at a given institution may be aggregated to form a single regional UDDI registry. Such a registry may not replicate the content of each UDDI node but may provide references to content providers.

UDDI is primarily meant as a registry for business services, and data structures within UDDI reflect this aspect. Information about a business that can be stored in UDDI may be classified as follows:

■ White pages—These contain basic contact information and identifiers about a company, including business name, address, contact information, and unique identifiers such as its Dun-and-Bradstreet (DUNS) numbers or tax IDs. This information allows others to discover Web Service based on business identification. In the context of Grid Computing, white pages can provide the retrieval of an IP address or the amount of memory available on a particular resource.

■ Yellow pages—These contain information that describes a Web Service using different business categories (taxonomies). This information allows others to

---

[27] [OASIS-UDDI] OASIS UDDI Specification Technical Committee. See documents at: *http://www.oasis-open.org/committees/tc_home.php?wg_abbrev=uddi-spec*

discover Web Services based on its categorization (such as flower sellers or car sellers).

■ Green pages—These contain technical information about Web Services that are exposed by a business, including references to specifications of interfaces, as well as support for pointers to various file and URL-based discovery mechanisms.

The data structures in UDDI are primarily meant to support business services and do not provide an effective representation of needs of the science and engineering community. Generally, these data structures allow encoding of properties related to white, yellow, and green pages defined above. These are therefore not adequate on their own for representing services in scientific disciplines, as no similar categorization of scientific areas is currently in use. There is also no unique naming scheme (such as DUNS/tax IDs) for representing providers of science or engineering services. Ongoing efforts at the Global Grid Forum seek to address some of these concerns.

10.8 UDDI AND OGSA

The UDDI OGSA bridge is illustrated in Figure 10.5, based on Colgrave.[28] To bootstrap the system, a UDDI registry may contain references to a HandleResolver, a Factory, or a ServiceGroup, for instance. The information maintained in UDDI could therefore be used to create a new instance of a service (this is based on the assuming that WSDL descriptions of services are registered with UDDI and SOAP/HTTP binding is required. Interaction between a Globus Toolkit environment and UDDI would be mediated via the OGSA UDDI bridge, enabling the description of Grid Services with a particular lifetime (achieved by defining a ServiceGroup that conforms to these properties). A service could join one or more of such groups via the ServiceGroupRegistration interface. A UDDI-aware HandleResolver would be used to identify service descriptions which have been registered within UDDI. There is no security support at present for authenticating users with the UDDI registry, although this is a significant requirement for business services to interact. The general consensus is that public UDDI registries should contain information that is universally accessibly. Use of X509 based digital certificates via the Grid Security Interface (GSI) is a useful way to provide secure access to the UDDI registry whereby the UDDI OGSA bridge verifies user credentials before accepting a query.

---

[28] [Colgrave03] John Colgrave, "UDDI and OGSI," presented at *Grid Information Services Workshop*, Edinburgh, UK, April 25, 2003.

The UDDI OGSA bridge is therefore meant to connect two environments without necessitating a major change in either. It is a useful compromise that can still enable the use of existing UDDI implementations by relating key structures in grid environments (such as Globus) with the UDDI registry.

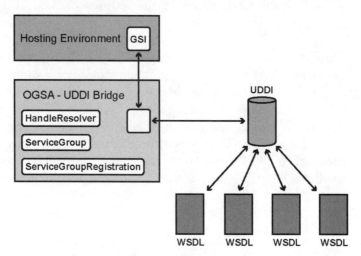

FIGURE 10.5   UDDI OGSA bridge.

10.9 UDDIE: UDDI EXTENSIONS AND IMPLEMENTATION

The approach adopted in UDDIe is different from the approach taken in the UDDI OGSA bridge, in that additional functionality has been added to UDDI to enable its use with Grid Services. This functionality also makes UDDIe a generic registry service for Web Services. UDDI may also find use in other applications. Information within UDDIe is also stored as a WSDL document but can make more effective use of the extensibility elements supported in OGSA. A set of user-defined attributes can now be recorded as part of a propertyBag in UDDIe, thereby allowing a user to discover services based on user-defined attributes.

UDDIe uses an XML schema that extends that used in standard UDDI[29] by using the same specification and standards for the registry data structures and application programming interface (API) specification for inquiring and publishing services. Extensions in UDDIe are based on four types of information: business information, service information, binding information, and information about

---

[29] [CoGmanual] G. von Laszewski, B. Alunkal, K. Amin, J. Gawor, M. Hategan, and S. Nijsure, "The Java CoG Kit User Manual," Argonne National Laboratory, Mathematics and Computer Science Division, 9700 S. Cass Ave, Argonne, IL 60439, U.S.A., MCS Technical Memorandum ANL/MCS-TM-259, Mar. 14, 2003.

specifications for services. A service may be discovered by sending requests based on service information. The extensions provided in UDDIe consist of the following:

■ Service Leasing—Service providers may want to make their service available for limited time periods (for security reasons, for instance) or the service may change often. Existing UDDI implementations lack support for services that change often or for unreliable hosting environments that may become un-accessible. To overcome this situation, UDDIe supports *finite* and *infinite* leases. A finite lease can be immediate or specified as a future lease. When using finite leases, service providers must define the exact period for which the service should be made available for discovery in the registry. The lease period is re-stricted by the maximum allowable lease period defined by the UDDIe admin-istrator. For example, if a service provider is interested in publishing a service in UDDIe for two hours, but the maximum granted lease is one hour, publica-tion of the service will be rejected by the registry. A *future* lease allows a service provider to make the lease period start at a future time; the service will be dis-coverable only after this lease has been activated. The future lease concept is supported to enable a systems administrator to directly encode dependencies in service invocation.

Alternatively, service providers may want to publish their services for an infinite period of time. Such leases are allowed in UDDIe, but only if the ratio of finite/infinite lease services is within a threshold (a parameter set by the UDDIe administrator). Effectively, an infinite lease is used to define persistent services, which must be maintained in UDDIe for long periods of time. These may include system critical services, which need to be held in the registry for a grid system to work effectively; examples are the Monitoring and Discovery Service (MDS) or various job managers that may be available. Conversely, every user-defined service must be granted a finite lease and would require a user to renew this periodically.

■ Replication—The UDDI Universal Business Registry (UBR)—illustrated in Figure 10.6—is conceptually a single system built from a group of UDDI nodes that have their data synchronized through replication. The UBR is intended to be a public registry that contains a number of different UDDI services. A series of operator UDDI nodes each host a copy of the content, thereby replicating content among one another. Content may be added to the UBR at a single node, and that operator node becomes the content master. Any subsequent updates or deletes of the data must occur at the operator node where the data were inserted. UDDIe can be used as a private operator node that is not part of the UBR. Private nodes do not have data synchronized with the UBR, so the in-formation contained within is distinct. The availability of private nodes is sig-nificant if an organization considers sharing their service content a security problem, for instance if a company does not want to expose certain service offerings and business processes to others. It is not clear who should own and manage the "topmost" UDDI nodes in the context of Grid Computing. Currently, major UDDI vendors such as IBM and HP are expected to own the topmost UDDI registry for business services; however, it is unclear how such a structure should be managed for Grid Services.

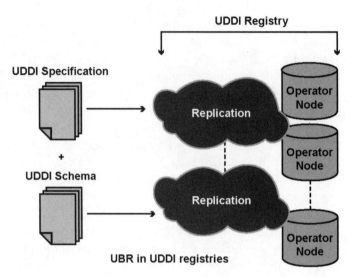

FIGURE 10.6 Replication UDDI nodes to form the Universal Business Registry.

In UDDIe, a businessService[30] structure represents a logical service and is the logical child of a businessEntity the provider of the service. Service properties are contained in the propertyBag entities such as the Quality of Service (QoS) that a service can provide or the methods available within a service that can be called by other services (an important feature missing in current UDDI implementations). Figure 10.7 illustrates the attributes associated with a property and consists of a propertyName, propertyType, and propertyValue. Some of these are user-defined attributes such as propertyType and can be number, string, method. Range-based checks, for instance, are allowed only if the propertyType is a number.

---

[30] [CoGmanual] G. von Laszewski, B. Alunkal, K. Amin, J. Gawor, M. Hategan, and S. Nijsure, "The Java CoG Kit User Manual," Argonne National Laboratory, Mathematics and Computer Science Division, 9700 S. Cass Ave, Argonne, IL 60439, U.S.A., MCS Technical Memorandum ANL/MCS-TM-259, Mar. 14, 2003.

An example of using `PropertyBag` is illustrated in code Listing 10.2. The data types that can be supported in the `PropertyType` may be extended by using xsd (XML Schema Description) tags and a user-defined namespace. The `propertyFind-Qualifier` is part of the query, whereas the other aspects of the `propertyBag` are also used to define particular properties.

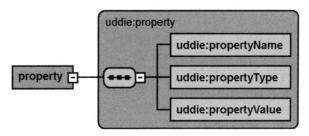

FIGURE 10.7 Property attributes.

Listing 10.2

```
<propertyBag>
 <property>
  <propertyFindQualifier>GREATER_THAN</propertyFindQualifier>
  <propertyName>CPU</propertyName>
  <propertyType>number</propertyType>
 <propertyValue>800</propertyValue>
 </property> ...
</propertyBag>
```

The API for interacting with the registry system extends three classes within existing UDDI implementations. The extensions provided in the API include the following:

- `saveService`: This set of APIs is used mainly for publishing service details. This has been extended from the original UDDI system to introduce dynamic metadata for services. Such metadata could be used to represents attributes such as cost of access, performance characteristics, or usage index associated with a service, along with information related to how a service is to be accessed and what parameters the service will return.
- `findService`: This set of APIs is used mainly for inquiry. In particular, we extend this set of APIs from the original UDDI to include queries based on

various information associatcd with services, such as Service Property and Service leasing.

- `getServiceDetails`: This set of APIs is used mainly for requesting more detailed information about services, such as BusinessKey information. We extend this set of APIs to include Service Property information.

- `renewLease`: This set of APIs is used both by the operator/UDDI administrator to control leasing information and by the service provider (SP) to renew and set leasing information. The leasing concept works as follows. Lease information must be associated with every service, either for a limited duration or for an infinite time period. The maximum number of infinite leased services is controlled by the operator to efficiently maintain the registry. In the case of limited duration, the SP provides a start-from date and an expiration date for the lease period. The operator has control on setting up the default leasing period. Moreover, if a lease expires, the SP could always renew the expired lease, provided that the request falls within the allowed number of times to renew a particular lease, which is controlled by the operator/UDDI administrator. When the lease period expires, the service becomes invalid and clients cannot make use of the expired services. It is important to continually renew a lease or request an infinite lease, and an event manager is used to alert all connected users if the lease of the service is about to expire.

- `startLeaseManager`: This newly defined set of APIs is used to monitor lease constraints by generating processes to monitor and update the lease period. A service with an expired lease is removed from the registry. The operator has control over how often to run these processes. The lease manager is started when the UDDIe registry is first initialized and must run as a background process. The lease manager must be configured by the UDDIe administrator to support a particular ratio of services with infinite and finite leases.

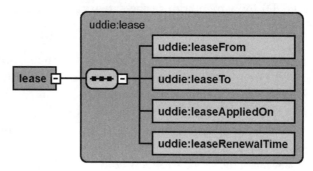

FIGURE 10.8 Lease element.

In addition to these sets of APIs, we introduce support for a *qualifier-based* search—to find services based on some property along with a qualifier value, such as LESS_THAN, GREATER_OR_EQUAL, EQUAL_TO, NOT_EQUAL_TO and GREATER_THAN. This set of qualifiers may be extended by the user. Support for logical operations is introduced to enable querying for properties with logical AND/OR operators. The search for services that match a particular request is based on the following set of operations.

```
element result set:
        A set containing all service keys that match the value of the
element
total result set:
        A set that contains all the element sets

FOR EACH element in the find_service message DO
  Fetch the services which matches the element value
  Add the services' key into the element result set
  Add the element result set into the total result set
END For Loop

IF Logical OR is required THEN
   final result set = Union of all element result set in the total
result set

IF Logical AND is required THEN
   final result set = Intersection of all element result set in the
total result set
```

UDDIe also provides support for error checking, and includes support for the following types of errors:

- Range checks—This check is used to ensure that the range specified for a particular property does not exceed its limits.
- Existence—This error check is used to ensure that a particular property actually exists, and is returned to the user via a DispositionReport.

```
<dispositionReport xmlns="urn:uddi-org:api_v3">
  <result errorno="***">
    <errInfo errCode="***"/></result></dispositionReport>
```

■ Duplicates—This check is used to ensure that a particular service does not have two definitions for the same PropertyType or that a particular property is not replicated. This is achieved by using the Unique tag.

■ Lease checking—A number of checks are provided to support leasing (soft state), such as InvalidLeaseDateException (a data quality check to ensure that the lease data conform to the required format), a RenewalTimeExceededException (a rule-based check to ensure that a particular lease has not been renewed more than a given number of times), and InfiniteLeaseOutofBoundException (a rule-based check to ensure that the number of Infinite to Finite leases does not exceed the threshold set by the systems administrator).

These extensions to the UDDI registry and query mechanisms provide flexibility in undertaking search, thereby making UDDI a more powerful search engine. The ability for UDDIe to co-exist with standard UDDI version is also important since it allows co-existence with existing UDDI deployments. Listing 10.3 illustrates a mathematical service with service properties added, and to be used within UDDIe. The description of the mathematical service illustrates the use of SOAP bindings for performing a particular request, and the use of QoS attributes of a service (as its properties). The QoS tag indicates that this service requires a CPU rating of 100, and a disk storage rating of 150 to execute successfully. Hence, the WSDL interface for the service indicates operations supported by the service, mechanisms for connecting to the service, and performance characteristics that need to be adhered to if the service is to run successfully.

Listing 10.3

```
<?xml version="1.0" encoding="UTF-8"?>
<wsdl:definitions xmlns:wsdl="http://schemas.xmlsoap.org/wsdl/"
xmlns="http://schemas.xmlsoap.org/wsdl/"
    xmlns:SOAP-ENC="http://schemas.xmlsoap.org/soap/encoding/"
xmlns:impl="http://MathService"
    xmlns:intf="http://MathService-Interface"
xmlns:tns1="http://DefaultNamespace"
xmlns:wsdlsoap="http://schemas.xmlsoap.org/wsdl/soap/"
xmlns:xsd="http://www.w3.org/2001/XMLSchema"
targetNamespace="http://MathService-Interface">
```

```
  <wsdl:message name="getSumResponse">
    <wsdl:part name="return" type="SOAP-ENC:string"/>
  </wsdl:message>
  <wsdl:message name="getSumRequest">
    <wsdl:part name="symbol" type="SOAP-ENC:string"/>
  </wsdl:message>

  <wsdl:portType name="MathService">
    <wsdl:operation name="getSum" parameterOrder="symbol">
      <wsdl:input message="intf:getSumRequest"/>
      <wsdl:output message="intf:getSumResponse"/>
    </wsdl:operation>
  </wsdl:portType>


wsdl:binding name="mathSoapBinding" type="intf:MathService">
    <wsdlsoap:binding style="rpc"
transport="http://schemas.xmlsoap.org/soap/http"/>
    <wsdl:operation name="getSum">
      <wsdlsoap:operation soapAction="" style="rpc"/>
      <wsdl:input>
        <wsdlsoap:body
encodingStyle="http://schemas.xmlsoap.org/soap/encoding/"
namespace="http://MathService-Interface" use="encoded"/>
      </wsdl:input>
      <wsdl:output>
          <wsdlsoap:body
encodingStyle="http://schemas.xmlsoap.org/soap/encoding/"
namespace="http://MathService-Interface" use="encoded"/>
      </wsdl:output>
    </wsdl:operation>
  </wsdl:binding>

  <wsdl:QoS>
   <cpu_count>100</cpu_count>
   <disk_storage>150</disk_storage>
  </wsdl:QoS>
</wsdl:definitions>
```

10.10 USES

UDDIe can be used in any applications where providers register their service for a limited duration or where users search for services based on range-based/logical criteria. Such criteria include business providers who want to register their services with attributes that are numeric, such as service cost, or service quantity, or when specifying performance or quality attributes associated with a service. Service registration for limited duration is particularly useful when the environment within which such a service is used is dynamic and may change often. Here, an infinite lease for a service may not be suitable or a service provider may predict a future time when the service may be usefully deployable and opt for a future lease. To use UDDIe, a user must define service properties, as illustrated in Listing 10.4. A UDDIe proxy is established to enable a user to publish and interrogate the registry. A service called Math is then registered, and properties associated with the service are specified. These properties are then registered into a propertyBag.

Listing 10.4

```
UDDIeProxy proxy = new UDDIeProxy();
proxy.setInquiryURL("http://localhost:8080/uddie/inquiry");
proxy.setPublishURL("http://localhost:8080/uddie/publish");

//Get Authorization by sending a username and password
// for the owner of the business

AuthToken token = proxy.get_authToken("ali", "ali");

//Define service name and add them to a vector
//The maximum allowed names is 5

Name name = new Name("Maths");
Vector names = new Vector();
names.add(name);

// Define Service propertiesProperty

property= new Property("CPU", "number", "50");
property.setPropertyFindQualifer(PropertyFindQualifiers.GREATER_THAN;
Property property2 =   new Property("RAM", "number" , "30");
property2.setPropertyFindQualifer(PropertyFindQualifiers.EQUAL_TO);
```

```
Vector properties = new Vector();properties.add(property);
properties.add(property2);
PropertyBag bag = new PropertyBag();
bag.setPropertyVector(properties);
```

We then define a FindQualifier,which may be used to search for the registered service and issue a query:

Listing 10.5

```
// Define Find Qualifer for property exact match (Logical AND)
FindQualifier findQualifier = newFindQualifier("exactPropertyMatch");
FindQualifier findQualifier2 = newFindQualifier("exactNameMatch");
FindQualifiers qualifiers = new FindQualifiers();
Vector qualifiersVector = new Vector();
qualifiersVector.add(findQualifier);
qualifiersVector.add(findQualifier2);
qualifiers.setFindQualifierVector(qualifiersVector);

// Send the query
ServiceList list = proxy.find_eService(null, names, null, bag, null,
qualifiers , 5);
```

Once a query has been issued, details of the services that match the requests are printed using a for loop, and returned in the ServicesInfos object, as illustrated in Listing 10.6. The first loop (using index i) returns a list of services, and the second (using index j) returns properties associated with each service.

Listing 10.6

```
ServiceInfos infos =   list.getServiceInfos();
Vector services =   infos.getServiceInfoVector();

for ( int i=0;i<services.size();i++){
    ServiceInfo service =   (ServiceInfo)services.get(i);
    System.out.println("Service  returned Name: " +
service.getName().getText());System.out.println("Service  returned Key:
" + service.getServiceKey());
eServiceDetail serviceDetail =proxy.get_eServiceDetail(
service.getServiceKey());
```

```
    Vector serviceVector =serviceDetail.getBusinessServiceVector();
    BusinessService returnedService
=(BusinessService)serviceVector.firstElement();
    Lease lease =  returnedService.getLease();System.out.println("lease
expiration date: " +lease.getLeaseExpirationDate());
    PropertyBag bag2 =  returnedService.getPropertyBag();
    Vector propertiesFound =  bag2.getPropertyVector();

for (intj=0;j<propertiesFound.size();j++){
    Property propertyFound = (Property)propertiesFound.get(j);
    System.out.println("Property Name: "
+propertyFound.getPropertyName());
    System.out.println("Property Value: "
+propertyFound.getPropertyValue());
    }
```

Figure 10.9 illustrates how a client can use the UDDIe registry. A request is received from an external client using HTTP and parsed by the Servlet. A client must distinguish whether the request is a UDDI request or a UDDIe request when publishing information about a service. However, when searching for a service, both UDDI and UDDIe requests are treated identically, as UDDIe extends the `find_-service` in UDDI with additional attributes. If these extensions are not implemented (as happens in existing UDDI registries), the default version of

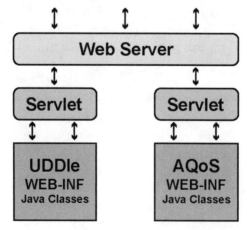

FIGURE 10.9 When using UDDIe, a client sends a request as an XML message to the Servlet, which parses the request and forwards it to the registry.

`find_service` will be invoked. If service properties are also specified in the call, however, then the updated version in UDDIe will be used. Java classes to implement UDDIe extensions are stored as a jar file and can be invoked by the parser when the extensions are encountered. Publishing service information (as illustrated in code fragments 5 and 6) requires the user to define properties associated with a service, and add these to the `propertyBag`.

10.11 QUALITY OF SERVICE MANAGEMENT

UDDIe can be used to support quality of service management (QoSM) in the context of Grid Computing. A service in this case represents either a scientific or engineering code or a mathematical library, and each service has a description provided in a WSDL document. A service also has other attributes, such as bandwidth, CPU, and memory requirements encoded in the service interface. Some of these parameters, such as bandwidth, packet jitter, and loss, are measured by a *bandwidth broker*—(a network monitoring program that is used to derive these low-level metrics[31]). These attributes allow a service user to choose between services based on their QoS attributes, rather than just their functional properties. The system is implemented in the context of our G-QoSM framework,[32] whereby clients/applications send their requests for services with QoS properties to our QoS broker (Listing10.7 gives an example of the WSDL document used). The broker processes the request and submits the service request portion to UDDIe. The UDDIe registry replies with a list of services that support this particular query. Finally, the broker applies a selection algorithm to select the most appropriate service with respect to the client/application request and sends the result to the client/application. The selection algorithm that the broker uses is based on the weighted average (WA) concept. In this algorithm, we introduce the notion of a QoS importance level, whereby the client/application is requested to associate a level of importance, such as High, Medium, or Low, with every QoS attribute. Based on this QoS importance level and the value of a QoS attribute, the algorithm computes the WA for every returned service, and selects the service with the highest WA.

---

[31] [bb] B. Teitelbaum, S. Hares, L. Dunn, R. Neilson, R. Vishy Narayan, and F. Reichmeyer. "Internet2 QBone: Building a Testbed for Differentiated Services," *IEEE Network*, 13(5):8–16, September/October 1999.

[32] [gqosm] R. Al-Ali, O. Rana, D. Walker, S. Jha, and S. Sohail. "G-QoSM: Grid Service Discovery Using Qos Properties," *Computing and Informatics Journal, Special Issue on Grid Computing*, 21(5), 2002.

Listing 10.7

```
<?xml version="1.0" encoding="UTF-8"?> <wsdl:definitions
xmlns:wsdl="http://schemas.xmlsoap.org/wsdl/" ... ]
targetNamespace="http://MyService-Interface">
<wsdl:messagename="printNameResponse">
</wsdl:message> ...
<wsdl:QoS>
 <service_cost> 5 </service_cost>
 <network_bandwidth> 256K </network_bandwith>
<memory> 48MB </memory>
    ....</wsdl:QoS>
</wsdl:definitions>
```

10.12 CONCLUSION

Chapter 10 discussed the role of the service abstraction in Grid Computing, followed by various mechanisms about how such an information service may be used to publish and discover such services. It described the need to base such an information service on Web Services technology, such as UDDI, and outlined how such technology may be connected to existing grid middlware, such as the Globus Toolkit. The implementation of a registry for Web Services, which extends UDDI, is specified. It is particularly useful when storing service properties with range-based attributes and also enables search for services based on these properties. UDDIe and WSDL provide an important mechanism for specifying and deploying Web Services especially when extending a WSDL document with additional attributes such as service quality and performance data. Using UDDIe, one also can publish information about method calls (when wrapping object based implementations) available within a service (as its properties) and to subsequently search based on these method signatures. The extension through propertybag therefore allows a better way to make existing object-based systems available as services especially when used with tools such as SOAPswitch[33] and Java2WSDL.[34]

---

[33] [actional02] Actional—Web Services Management Solutions. "SOAPswitch." See Web site at *http://www.actional.com/products/soapswitch/index.asp*. Last visited: October 2002.

[34] [wstk] IBM—Alphaworks, "Web Services Toolkit." See Web site at *http://www.alphaworks.ibm.com/tech/webservicestoolkit/*. Last visited October 2002.

Download

UDDIe may be downloaded as an open-source software, protected via the GNU public license, from *http://www.cs.cf.ac.uk/User/A.Shaikhali/uddie/*. It extends the public domain implementation of UDDI available at uddi.org and consequently requires a number of additional programs to be available to run, such as uddi4j, a Web server that supports Java Applets (such as Tomcat), a relational database management system (such as Oracle or MSSQL), JDBC drivers for the particular database selected, and a Java virtual machine (shipped as part of the Java Development Kit). A precompiled version of UDDIe is also available at the above site and has been tested with Tomcat server 4.1, Oracle 9i, the Oracle Thin driver, and JDK 4.0. A Java client for UDDIe is also provided at this site and may be used as a standalone application or embedded into an existing Java program. An Applet implementation of the client is provided at the Web site to allow external users to experiment with UDDIe online.

Acknowledgments

We thank Rashid J. Al-Ali for his work on implementing quality of service properties using UDDIe, and members of the Welsh e-Science Centre.

The work of Dr. von Laszewski was supported by the Mathematical, Information, and Computational Science Division subprogram of the Office of Advanced Scientific Computing Research, Office of Science, U.S. Department of Energy, under Contract W-31-109-Eng-38. DARPA, DOE, and NSF support Globus Project research and development. The Java CoG Kit Project is supported by DOE SciDAC and NSF Alliance.

11

Desktop Supercomputing: Native Programming for Grids

Matt Oberdorfer and Jim Gutowski

Engineered Intelligence

In This Chapter

- Parallel Programming Paradigms
- Problems with Current Parallel Programming Paradigms
- Desktop Supercomputing with CxC

MNSP
Multiple Node Single Processor

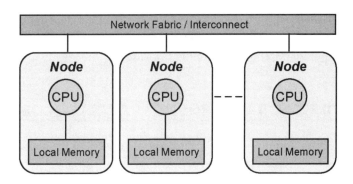

11.1 INTRODUCTION

The world of supercomputing is changing due to the emergence of new technologies and innovative approaches. Supercomputers still offer the highest performance for the highest price, but cluster computers, consisting of low-cost, high-performance processors connected with high-speed networks, offer almost comparable performance at a fraction of the cost. Pioneers in academia and some corporations have implemented supercomputing clusters; however, their use has been restricted because of the difficulty in programming these systems.

High-performance computing (HPC) with clusters offers great promise for meeting the needs of a broad community of engineers and scientists, but is limited by the complexities of developing software applications for these HPC systems. Parallel programming for supercomputers and HPC clusters requires knowledge of the OS; hardware and network configuration; message-passing and/or shared-memory programming approaches; and access to the supercomputer or cluster for programming, testing, and debugging. Until now, only a select group of experts were able realize the HPC benefits. Desktop Supercomputing changes the rules of the game. This chapter will looks at how it all fits together.

With *desktop supercomputing*, inventors can easily develop parallel algorithms and applications on their desktop systems; prototype, test, and compile the code; and run the executable unchanged on supercomputers or clusters or grids. The way it works is pretty simple: the development environment and operating system runs on any desktop, or even on a laptop. Desktop Supercomputing allows creating, prototyping, and running parallel algorithms anywhere and anytime offline—meaning without access to a supercomputer or supercomputing cluster. The executable is hardware independent and runs on a virtual parallel machine. When executed on a laptop or desktop, all parallel processors are simulated on the single processor of the physical desktop computer. Running the same unchanged executable, however, on a supercomputer, SMP, or on a supercomputing cluster or cluster grid shows parallel performance with almost linear speedup and scalability.

11.2 HISTORICAL BACKGROUND–PARALLEL COMPUTING

Traditional computer system hardware for parallel processing can be broadly divided into four categories, as defined by Flynn[1] 1966:

Historically, only two categories were of interest for parallel computing: SIMD, representing *synchronous parallelism*, and MIMD, representing *asynchronous par-*

---

[1] M. Flynn, "Very High Speed Computing Systems," *Proceedings of IEEE*, vol. 54, 1966, pp. 1901–1909.

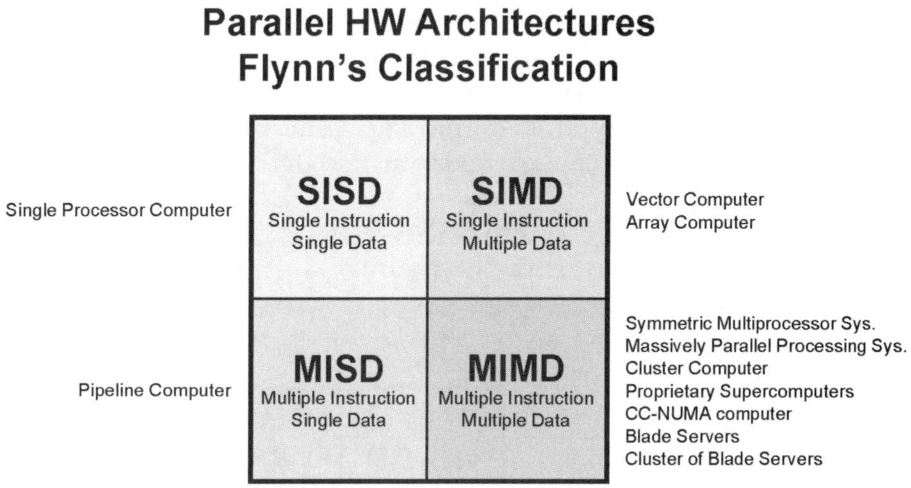

Parallel HW Architectures
Flynn's Classification

FIGURE 11.1 Flynn's classification.

allelism, as shown in Figure 11.1. In synchronous parallelism, there is only one control flow, meaning there is only one control processor working on the program orchestrating the execution of programs on all other processors. In asynchronous parallelism, there are many control flows, meaning each processor works independently but concurrently through its own program. In asynchronous computers, every data exchange needs to be synchronized. Both are described in detail below.

11.2.1 MIMD Computers

The language of desktop supercomputing, CxC, combines the advantages of C, Java, and Fortran and is designed for MIMD architectures that are considered here in more detail. The MIMD category was so broadly defined by Flynn that any parallel computer not following the SIMD approach (one-program-on-one-processor-controls-all-others) automatically fell into the MIMD category. Below is a sample list of popular MIMD hardware architectures:

- Symmetric Multiprocessing Systems (SMP)
- Massively Parallel Processing Systems (MPP)
- Cluster computers
- Proprietary supercomputers
- Cache-Coherent-Non-Uniform Memory Access (CC-NUMA) computers
- Blade servers
- Clusters of blade servers

It is reasonable and useful to introduce a more appropriate classification that differentiates between various MIMD architectures.

In order to do so, a definition is needed: A *node* is a standalone computer system that contains one or more processors that have access to the local memory of this system (closely coupled). Nodes can be connected by a network fabric to form

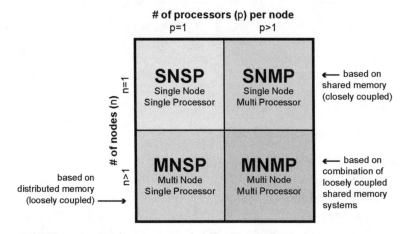

FIGURE 11.2   MIMD computer classification.

a cluster of compute systems where each can communicate to the others, sending messages back and forth over the network (loosely coupled).

The four categories (Figure 11.2) are divided based on the number of processors per node (one or more than one) and based on the number of loosely coupled/networked nodes (one or more than one).

- The category for Single-Node/Single-Processor (SNSP) computers, also known as von-Neumann computers, is the same as Flynn's Single-Instruction-Single-Data (SISD) category.
- Single-Node/Multiple-Processors (SNMP—Figure 11.3) computers are shared-memory computers having multiple processors within the same node accessing the same memory. Representatives are blade servers, symmetric multiprocessing systems (SMP), CC-NUMA architectures, and other custom-made high-performance computers. Array and vector computers (SIMD) would fall into this category.
- Multiple-Node/Single-Processor (MNSP—Figure 11.4) computers are distributed-memory computers represented by a network of workstations

SNMP
Single Node Multiple Processor

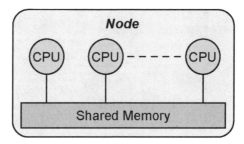

FIGURE 11.3 SNMP computer.

MNSP
Multiple Node Single Processor

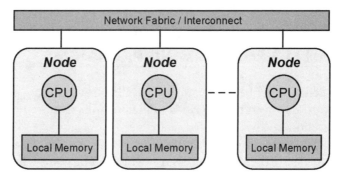

FIGURE 11.4 MNSP computer.

(NOW), a cluster of off-the-shelf components (COTS), or other distributed memory computers. Clusters of computers with two or more processors fall into this category if the parallel operating system treats each processor as a separate system. Basically, this covers computer systems where all processors are loosely coupled.

■ Multiple-Node/Multiple-Processor systems (MNMP—Figure 11.5) involve multiple shared-memory computers (SNMPs) connected by a network. MNMPs are a loosely coupled cluster of closely coupled nodes. Typical representatives for loosely couple shared-memory computers are SMP clusters or clusters of blade servers.

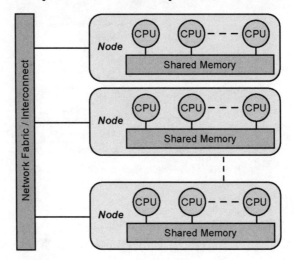

MNMP
Multiple Node Multiple Processor

FIGURE 11.5 MNMP computer.

11.3 PARALLEL PROGRAMMING PARADIGMS

This section describes the historical parallel programming paradigms used in the hardware architectures described in the previous section. It provides background information for the interested reader.

SNSPs preemptive multitasking is used as the parallel processing model. All processes share the same processor—which spends only a limited amount of time on each process—so their execution appears to be quasi-parallel. The local memory can usually be accessed by all threads/processes during their execution time.

Now consider SNMP (shared-memory parallel) computers, which are based on symmetric multiprocessing hardware architecture.

Shared memory is used as the parallel processing model on a SNMP computer with multiple processors in the same node. Here, each processor works on processes truly in parallel and each process can access the shared memory of the compute node. Typically connected through a high-speed connection fabric, they represent this single shared-memory system (Fig. 11.7). The operating system has been parallelized so that each processor can access the system memory at the same time. The shared-memory programming model is easy to use since all processors are able to run a partitioned version of sequential algorithms created for single processor sys-

Parallel Programming Paradigms

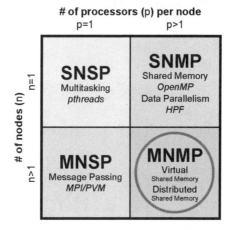

FIGURE 11.6 Parallel programming paradigms for parallel hardware.

tems. One disadvantage of SNMP systems is that *scalability* is limited to a small number of processors. Limitations are based on system design and include problems such as bottlenecks with the memory connection fabric or I/O channels, as every memory and I/O request has to go through the connection or I/O fabric. This is true whether the data being requested are shared by multiple processors or used only by one processor. Known methods for shared-memory (asynchronous) parallelism are OpenMP, Linda, or Global Arrays (GA). Data parallel synchronous parallelism has been established with High Performance FORTRAN (HPF).

Shared Memory Paradigm
Symmetric Multi Processing

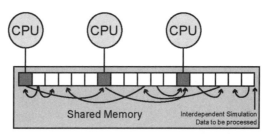

FIGURE 11.7 Data access in SNMP computer.

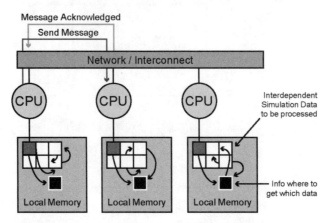

FIGURE 11.8 Data access in MNSP computers.

The programming model for standard MNSP (distributed-memory) computers, such as clusters and MPP systems, usually involves a *message-passing model* (Fig. 11.8). In a message-passing model, parallel programs must explicitly specify communication functions on the sender and the receiver sides. When data needed in a computation are not present on the local computer, data transfer is performed by issuing a send function to the remote computer holding the data and then issuing a receive function at the local computer. The process for passing information from one computer to another computer via the network includes data transfer from a running application to a device driver; the device driver then assembles a message to be transferred into packets to the remote computer, which is subsequently sent through networks and cables to the receiving computer. On the receiving computer's side, the mirrored receiving process has to be initiated: the application triggers "wait for receiving a message," using the device driver. Finally, the message arrives in packets and is reconstructed by the device driver, to be handed over to the waiting application. Disadvantages for the message-passing model include time delay/loss due to waiting on both transmitting and receiving ends; synchronization problems, such as deadlocks (applications can wait indefinitely if a sender sends data to a remote computer not ready to receive it); data loss, as each sender needs a complementing receiver (if one fails, data gets lost); and, finally, difficult programming—algorithms have to be specifically programmed for the message-passing model and sequential algorithms cannot be reused without significant changes.

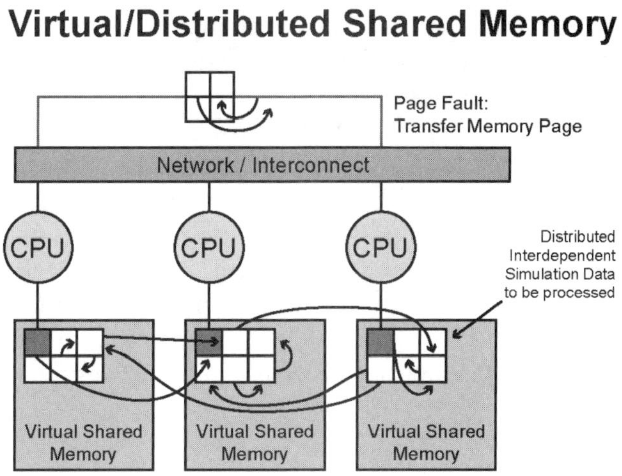

Virtual/Distributed Shared Memory

FIGURE 11.9   Data access in MNSP computer.

An approach used to overcome the difficult-to-program problem of the message-passing model is simulation of the shared-memory model on top of a distributed-memory environment. This is called *Distributed Shared Memory (DSM)* or Virtual Shared Memory and provides the function of *shared memory*, even though physical memory is distributed among nodes in the system. The disadvantage is loss of performance: every time a processor tries to access data in a remote computer, the local computer performs message passing of a whole memory page. This leads to huge network traffic and the network becomes such a significant bottleneck that the decreased performance becomes unacceptable for most applications.

11.4 PROBLEMS OF CURRENT PARALLEL PROGRAMMING PARADIGMS

Although parallel computers have been around for over two decades, programming them is still a new art for most software developers. In some cases, it is because parallel computers are very expensive and not available to most software developers. In recent years, cluster computers built from commodity products have been changing this, and making supercomputer power affordable, even for individuals.

A bigger challenge is the complexity of parallel programming. With different architectures, there are also different parallel programming paradigms, each having advantages and disadvantages. The small crowd of elite people with access to parallel computers even splits into multiple groups favoring different paradigms.

Additionally, there has been no satisfying model for Multiple-Node/Multiple-Processor computers, since simulated shared memory leads to unacceptable performance for the application.

In summary, the Shared-Memory Programming Model and the Message-Passing Programming Model offer both advantages and disadvantages. Algorithms implemented in one model have to be reprogrammed with significant changes to work in the other model. There is no effective parallel processing paradigm that works for both SNMP and MNSP systems. Mixing models is unacceptable due to complexity in the programs that makes maintenance difficult. Complexity of programming and lack of a standardized programming model for hybrid compute clusters, as described in Chapter 10, is an unsatisfying and unacceptable situation.

11.5 DESKTOP SUPERCOMPUTING: SOLVING THE PARALLEL PROGRAMMING PROBLEM

An innovative approach to solving the problems presented above is to split the development of the application from its execution on the physical hardware. With desktop supercomputing, the development environment runs on a desktop or laptop PC, and the scientist or engineer can program a virtual parallel computer—one that fits his problem, not the physical architecture of the SMP or other system. Using the parallel language CxC, scientists can create solutions to the scientific problem, not to the computer problems associated with shared-memory or message-passing paradigms. Using a distributed run-time environment to map the application to the hardware, desktop supercomputing frees the developer from concerns over the supercomputing platform he will use, its interconnection, operating system, and other architectural considerations. CxC desktop supercomputing

ON THE CD development system is included on the CD-ROM.

11.6 DESKTOP SUPERCOMPUTING PROGRAMMING PARADIGM

The key technology of Desktop Supercomputing is the *Connected Memory Paradigm* that allows developers to focus on building parallel algorithms by creating a virtual parallel computer consisting of virtual processing elements. It effectively maps, distributes, and executes programs on any available physical hardware. Then it maps a virtual parallel computer to available physical hardware, with creation of algorithms independent of any particular architecture. Therefore, Desktop Super-computing makes the parallelization process for even complex problems simple. It enables:

- Ease of programming—The language CxC allows developers to design algo-rithms by defining a virtual parallel computer—instead of having to fit algo-rithms into the boundaries and restriction of a real computer.
- Architecture Independence—Eexecutables run on any of the following archi-tectures without modification: SNMP, MNSP, or MNMP. Today developers can use shared memory on SNMP and message passing on MNSP architectures that are distinctly different, requiring significant effort to rewrite programs for the other architecture.
- Scalability—Developers can create programs on small computers and run these same programs on a cluster of hundreds or thousands of connected computers. This scalability allows testing of algorithms in a laboratory environment and tackling problems of sizes not previously solvable.
- Enhancement—It has the ability to unleash the performance of MNMP com-puters that have the best performance/price ratio of all parallel computers.

Desktop Supercomputing with CxC offers the advantages of message passing—using distributed-computing solutions—with the easier programmability of shared memory.

11.7 PARALLEL PROGRAMMING IN CXC

CxC is the language of Desktop Supercomputing. When using CxC, you undergo an intrinsic parallelization process by creating a virtual parallel computer that may consist of millions of parallel processing elements communicating to each other via topology. Every CxC program consists of three main creation steps:

- Specify the Virtual Parallel Computer Architecture.
- Define the Communication Topology.
- Implement the Parallel Programs.

In the first step, you create a virtual parallel computer architecture consisting of array controllers and parallel processing units (PPUs). You then define the communication topology between them in a second step. The third and final step is implementation of the programs running on each PPU. In the context of this document, "virtual" means the computer, and its components are created and represented in software and mapped later to the available hardware. Let's look at a simple example, which is a parallel version of the first program in Kernighan and Ritchie's classic test, The C Parallel Programming Language:

```
//// My first CxC program hello.cxc
//// controller and unit declaration

controller ArrayController    // create processor controller
 {                            // create parallel processors
  unit ParallelProcessor[30];
 }


//// don't need a topology since there is no communication
//// between processors taking place

//// program implementations

main hello(10)                // execute 10 times
 {
  program ArrayController     // program for all processors
   {                          // of declared controller
    println("hello parallel world!");
   }
 }
```

This simple CxC program creates 30 processors that will run all the same program. The parallel machine will be executed 10 times. The result is the following output 300 times (30 processors * 10 times executed).

```
hello parallel world!
hello parallel world!
hello parallel world!
 ...
```

And the good news is you can run it on your laptop in a couple of milliseconds. Imagine—if you increase the number of parallel processors from 30 to 30,000,000 you won't be able to run it on your laptop anymore. But on a supercomputer or cluster with 300 nodes it will run in 5 seconds.

11.8 PARALLELIZING EXISTING APPLICATIONS

CxC solution removes computational, platform, and scalability barriers; significantly lowers overall time and costs of development; and enables a new parallel computing paradigm, which provides the best platform for highly parallel applications.

CxC also offer an easy way to parallelize existing serial applications and works as an intermediate "glue" of FORTRAN, C, and C++ functions. Because CxC allows to call any existing C/C++ or FORTRAN functions it also maintains the orginal performance that have been achieved in these libraries.

CxC offers a great way to simplify the development and implementation of parallel algorithms with a huge number of interdependent elements and their interactions. Historically the simulation of interacting particles is considered as some the most complex and challenging applications that require tremendous computational resources. The sheer size of the mathematical problems, their very complexity combined with the required computational power is the reason why these kinds of applications fall into the category of the "Grand Challenges" as defined by the government.

11.9 CONCLUSION

Parallel computing will become more pervasive as more systems are developed with multiple processors, processors improve in performance, and connectivity reaches the speed of the memory bus in shared-memory supercomputers. Cluster computing is already the fastest growing segment in HPC and will replace traditional supercomputing in time. Intel's Hyperthreading is another path that emphasizes and validates that parallel computing will become a pervasive commodity. HPC Grids, Cluster Grids, and Desktop Grids are proliferating. Easy programming tools are required by the engineers and scientists who are trying to solve complex problems. Desktop supercomputing programming with CxC fulfills the promise of making large-scale computing systems available to every inventor.

12 Grid-Enabling Software Applications

Gord Watts and Steve Forde

GridIron Software, Inc.

In This Chapter

- Requirements for Grid-Enabling Software
- Grid Programming Tools: GridIron XLR8™
- Grid Enabling a MPEG4 Encoder—An Example

Nodes	Physical CPU	Logical CPU	No. of XLR8 Peers	Projected linear time (seconds)	Actual time (seconds)	% Deviation
1	1	2	2	8,820	8,820	0.00%
2	4	8	8	2,205	2,004	-10.03%
4	8	16	16	1,103	1,053	-4.70%
6	12	24	24	735	739	0.54%
8	16	32	32	551	615	10.37%
10	20	40	40	441	523	15.68%
12	24	48	48	367	472	22.14%

12.1 INTRODUCTION

To date, attention has almost exclusively focused on the *supply side* aspects of grid and utility computing. Most effort has been invested in developing *grid infrastructure* technologies: Middleware for the distribution and management of grid resources, hardware components such as rack-mounted servers and blade computers, and standards and protocols to ensure interoperability.

Although this grid infrastructure is essential, no broad adoption of grid computing, much less the realization of the "computing-as-utility" vision, will occur unless there is concomitant *demand* for the virtualized computing resources they provide. Specifically, there must be a sufficient number of useful, grid-enabled applications that can readily and easily take advantage of the grid.

This chapter explores the importance of making software applications grid-ready by identifying the practical business benefits that grid-enabled applications provide, discussing the necessary investments and required resources, and outlining the software modification process. A representative example of a mainstream software application modified for grid deployment illustrates the main discussion points.

12.2 GRID COMPUTING: DISCONTINUOUS INNOVATION OR MASSIVE YAWN?

Detractors of Grid Computing maintain that claims of the technology's importance and expected discontinuous impact are greatly exaggerated. Justifiably, they point to the fact that the failure to accommodate grid technology in the software development model bottlenecks all of the vast efforts to develop back-end grid infrastructure. The "if you build it, they will come" mentality of building out of supply side infrastructure patently ignores the demand side view.

Gartner's "hype cycle" tool for technology assessment posits that there is an inverse relationship between the attention a technology receives and its level of usage. Grid Computing technology, and the overlying notion of the utility computing model, have clearly reached the "Peak of Inflated Expectations" that Gartner says is characterized by a "frenzy of publicity" that "typically generates over-enthusiasm and unrealistic expectations."[1] The attention currently lavished on Grid Computing by media and analysts and its endorsement by technology behemoths such as IBM, Hewlett-Packard®, Sun Microsystems, and Intel mask the fact that few people are implementing the technology at present.

---

[1] Gartner, Inc., *http://www3.gartner.com/pages/story.php.id.8795.s.8.jsp*

If it has not already begun the slide into the abyss, Grid Computing is poised to descend into the "Trough of Disillusionment" by failing to deliver on its much-ballyhooed potential. The salient question is whether grid computing will languish and fade from memory, or whether sufficient practical uses and benefits will emerge to move the technology up the "Slope of Enlightenment" and attain the broad adoption required to reach the "Plateau of Productivity."

For grids to be broadly adopted and for the utility computing model to be viable, attention must rapidly shift to the practical needs of the ultimate end users of the technology. Entirely excluded from the grid discussion are the voices of the people who will "plug in" to this so-called compute utility, likened to that of a power grid. To follow the utility metaphor, everyone is worried about power generation (harnessing compute resources), transmission and distribution (systems for delivery and management), and wiring households with outlets (standards and protocols); nobody is thinking about how the appliances (applications) will need to be modified to plug in to the outlets nor the profound impact this will have on the way consumers (users) boil water and make toast.

The touch point connecting users with the grid is the applications they use. The degree to which applications are able to deliver the benefits of the grid to their users while simultaneously minimizing the amount of discomfort they experience in doing so will ultimately determine whether grid technology rises up the slope.

12.3 THE NEEDS OF GRID USERS

In understanding the needs of those on the demand side, there are three groups of stakeholders to consider: Application end users, business enterprises, and application developers.

The paramount concern of application *end users* is whether the benefits of the grid outweigh the difficulty they must incur in order to use it. The primary need of end users is that grid-enabled applications be *simple to use*. Adoption of the grid by end users will only occur if they do not have to fundamentally change the way they use an application. Furthermore, application end users will be intolerant of any requirement for extensive configuration or management. A zero configuration, zero administration "plug and play" approach is de rigueur for end users.

Business enterprises represent a special class of aggregated end users that have additional and different concerns. Like end users, enterprises will require the benefits of the grid to outweigh the difficulty they must incur in order to use it; however, in the case of the business enterprise this will be determined by calculating the return on investment that grid implementation provides.

Although some business costs associated with grid deployment are straightforward, such as the cost of grid infrastructure and project-based costs for implementation, others are less tangible. One of the major "soft" costs of grid deployment is the expense associated with business process engineering, or the fundamental changes that must be made in business processes in order to derive benefits from a compute grid. For this reason, the business enterprise shares the end user need for simplicity of use. The enterprise, however, has the additional concern of control and management. With many users and many resources to manage according to various business rules and shifting priorities, the business enterprise additionally needs the grid to be *simple to manage.*

Finally, the onus is entirely on *application developers* to do all the heavy lifting required to grid-enable software applications. The burden of modifying software application source code in order to allow distributed computing on a grid falls on the shoulders of independent software vendors (ISVs), in house developers, and third-party solutions integrators (SIs). The way applications are made will need to change significantly in order to fully leverage the capabilities of the grid, and perfectly good existing code will require modification. Like end users and the business enterprise, the benefits of the grid must outweigh the difficulty the developer must incur in creating new applications or changing existing applications. This will require grid capability to be a software *feature* that offers enough recognized benefit to application end users and business enterprises to warrant the development effort required. For the application developer, it must be *simple to develop* grid-enabled applications.

12.4 GRID DEPLOYMENT CRITERIA

To determine if it is worthwhile to deploy a compute grid—whether local, enterprise-wide, or global in scope—it is essential to establish whether or not there is sufficient resulting benefit or return derived from the required investment of effort, time, money, and resources.

There are three significant benefits that can be achieved with Grid Computing. First, the grid is capable of providing powerful processing capacity that meets the extreme requirements of high performance computing applications. Secondly, Grid Computing allows computationally intensive software applications to run significantly faster. Finally, grids can raise the efficiency of computing resources in an enterprise network from the typical 10 percent usage of desktops and 30 percent utilization of server capacity to the 80 percent to 90 percent range.

For enterprises that have intense, mission-critical computing requirements or that have computing initiatives that are highly strategic, grids are being used to provide powerful computing capability that is either an alternative or a supplement to expensive supercomputing hardware. Gene sequencing and protein folding are good examples of high performance applications that are mission critical for genomics and proteomics companies, and vital to the drug discovery industry. Forecasting and predictive modeling are examples of highly strategic computing applications that provide marketplace advantage in the hyper-competitive financial services, oil, and gas industries.

The tremendous speed that can be achieved with even a modest grid can yield significant benefit. Faster applications mean time to market and productivity advantages for businesses. Software vendors who provide faster applications offer their customers the single most desired end-user feature—speed. A grid that delivers linear or near-linear application speed up (i.e., an application runs nearly twice as fast on two identical machines, three times as fast on three machines, etc.) means that a grid comprised of just two or three computers results in a time-savings of 200 percent to 300 percent.

One of the most compelling benefits of Grid Computing to enterprises is the potential to more efficiently use computing resources. Increasing utilization to the 80 percent to 90 percent range represents a threefold improvement for the average server and almost tenfold improvement for the average desktop computer. Increased utilization directly translates to reduced capital expenditure and maintenance costs.

Although the business benefits of grid and utility computing are lucrative, there is a caveat. The vast majority of current grid implementations are for high performance computing. Although these earliest adopters are able to substitute grids and compute clusters for expensive high-end computing hardware at a fraction of a cost, few high performance applications are of relevance to most enterprises, and virtually none are practical for the SME (small- to medium size enterprise), prosumer or consumer markets. There is nothing utilitarian about the use of grids for this elite class of high performance computer software.

To realize the broader business benefits associated with increased application speed and improved resource utilization, there must be a sufficient number of applications able to take advantage of the computing resources that the grid provides. Even though it may be possible to increase the *availability* of a computing resource to 80 percent or 90 percent, it will not actually be *utilized* unless there are applications to fill that capacity.

The market for Grid Computing will remain small and there will be no adoption of the utility computing model unless demand expands beyond a limited number of high performance computing applications. Without a sufficient quantity of useful, grid-enabled applications, Grid Computing will remain merely a novelty.

12.5 METHODS OF GRID DEPLOYMENT

Clearly, the determination of whether or not to implement a compute grid, and ultimately, the demand for utility computing as a whole, is largely dependent on whether or not there are sufficient applications to take advantage of the virtualized compute resources.

Although grid deployment is simply not warranted for many applications, those that are *computationally intensive* in nature are particularly likely to benefit from grids. Computationally intensive means that an application either requires large amounts of processing or handles large amounts of data (or both).

Although computationally intensive is often taken as analogous to high performance, this is an unnecessarily limiting as well as false assumption. Computationally intensive can more broadly refer to any application where a user must wait an unacceptable length of time for processing or data handling to occur on a single computer.

There are two different methods for grid deployment of applications that are processing intensive: The *scripted batch queue* distribution method, and the *programmatic* method of coding for parallel distributed processing.

Distributed Resource Management (DRM) solutions, such as Sun ONE Grid Engine and Platform LSF use the batch queue method. With the batch queue approach, grid deployment can be achieved with little or no code modification by replicating an application across several computers and distributing computing jobs through scripting. Multiple jobs from application users are submitted and queued, and the DRM software most efficiently and appropriately allocates the jobs to the available computing resources. A batch queue implementation, for example, would allow ten jobs that each require one hour to process to all be completed in one hour when distributed across a grid of ten identical computers. Conversely, processing the same ten jobs serially on a single computer would require ten hours.

For the batch queue method to be effective, the following conditions must exist: There is a large quantity of jobs for a single application to process; there is a large pool of computing resources available and capable of performing processing; the application is scriptable; and there is sufficient MIS/IT expertise to configure, deploy, and manage the batch queue middleware and scripting.

Without a large quantity of jobs or without a large pool of resources, the effectiveness of the batch queue approach is diminished or even nullified. If there is only one job to be processed, for example, or if there is only one computer to do the processing, the time required to complete the job(s) cannot be reduced.

The programmatic method for parallel distributing processing entails modifying software source code to allow a single, large job to be broken down into several smaller tasks that can each be sent out and processed individually on separate computers and the results returned to be recompiled. Instead of requiring the application to be installed, each task provides the compute resource with an instruction set and only the portion of data it requires for the computation.

Unlike the batch queue method, the programmatic method can be effective with even a single job, as few as two computing resources, and can be applied to both scriptable and non-scriptable applications. For example, a single job that requires one hour to process could be completed in thirty minutes on two computers, and in six minutes across a grid of ten identical computers.

Although the programmatic method does not require DRM middleware or IT expertise to implement and manage, it does require software development expertise and access to application source code.

12.6 WHEN TO GRID-ENABLE SOFTWARE

Once applications have been determined to be computationally intensive and categorized as either scriptable or non-scriptable, there comes the task of determining the most applicable method of grid deployment.

The case of non-scriptable applications is straightforward. Non-scriptable applications can only be grid processed by means of the programmatic method. Scriptable applications, however, can be deployed using either the batch queue or programmatic methods.

Three key factors to consider in determining the most appropriate method of grid deploying a scriptable application are: The number of jobs to be processed; the number of available resources; and the size of the jobs and capacity of the resources.

Because batch queue solutions match the jobs to the available resources, they perform optimally the more closely parity is achieved between the number of jobs to be processed and the quantity and capacity of available computing resources. Batch queue is most beneficial when there is a many-to-many mapping of jobs to resources, and when there is a high level of user control and resource management required.

In addition to quantity, the size of jobs and the capacity of resources are highly relevant. Only high capacity resources can process large jobs. If there is a mismatch between the capacity of the available resources and the size of the jobs, batch queue will be much less effective. Low capacity resources will be relatively underutilized while high capacity resources will be in constant demand.

A fairly constant and consistent stream of job submissions is also important. Ideally, there should always be one job in the queue waiting to be processed, and one added as its predecessor is delegated to the grid. Batch queue is least effective when job submission is highly sporadic, with bursts of large quantities of job submissions punctuated by long idle periods.

Some other important considerations: Because batch queue requires the application to be replicated on all of the computing resources, software deployment and the cost of licensing can potentially be a factor, particularly if the grid is comprised of a large number of compute resources. If several applications are deployed, memory requirements may also be an issue.

It may, therefore, be preferable to distribute scriptable applications using the programmatic method if:

- There are few jobs.
- There are few resources.
- Typical job size is very large.
- The capacity of most resources is small.
- User self sufficiency is preferred to IT control and management.
- The frequency of job submission is sporadic.
- Software deployment on each compute resource is prohibitive for reasons of cost, provisioning, or inadequate system requirements.

Generally, the batch queue method for distributing scriptable applications works well in large enterprise environments with many users, many resources, sufficient IT resources, and business rules and policies which dictate the appropriate use and priority of compute resources. The programmatic method generally works well for smaller enterprises with highly processing-intensive applications, or with grids made up of low capacity "commodity" computers. It is also the exclusive method of distributing non-scriptable applications.

Although distinct and different, batch queue and programmatic distribution methods can be highly complementary. By taking advantage of standards such as Distributed Resource Management Application API (DRMAA) and OGSA, programmatic solutions can interface with DRM software to create an integrated environment where both scriptable and non-scriptable applications can be distributed together on a common grid infrastructure and can be administered using a single management environment. This hybrid approach essentially delivers the best of both worlds: The ability to manage many users and resources that batch queue provides combined with the user autonomy and wider applicability of the programmatic method.

12.7 REQUIREMENTS FOR GRID-ENABLING SOFTWARE

Two requirements must be met in order to modify software for grid deployment: *Access* to the application source code and the ability to *modify* it; in other words, both the legal right and the development expertise necessary to change an uncompiled application. There are three groups that meet these requirements.

The first group consists of independent software vendors (ISVs) who develop and commercially distribute software applications. ISVs own their software code and have software developers in their employ.

The second group is made up of academic institutions and enterprises in research-intensive industries such as life sciences that use open source software applications. Open source software licenses permit modifications of code, and in many cases allow redistribution of the modified version, subject to certain conditions.

The third group consists of enterprises that have developed their own proprietary software applications for internal deployment, often with a view to securing competitive advantage through superior implementation of information technology. As these applications are proprietary, enterprises typically own or in some fashion retain intellectual property rights to the source code.

Both open source and proprietary software applications can be modified for grid deployment either by internal software developers or third-party solution integrators (SIs).

12.8 GRID PROGRAMMING TOOLS AND EXPERTISE

Although the potential business benefits of enabling software to take advantage of Grid Computing are lucrative and compelling, few applications have been modified. This is largely due to the fact that up until now the process of modifying source code for parallel distributed processing has been highly complex and non-trivial, often requiring changes to and the addition of many lines of code.

The primary tools for modifying code to enable parallel distributed processing have been protocols such as MPI (message passing interface) and PVM (parallel virtual machine). These protocols provide an exhaustive communications library that gives a developer all the programming calls required to programmatically manage the sending and receiving of data between processes. Although the protocols are very complete, they are also exceedingly vast. More calls are provided by the protocols than are provided by some operating systems.

In addition to the sheer size of the library provided by parallel programming protocols, the calls are very low level, generally requiring extensive modification to existing lines of code and the addition of much supplemental code. Application development is an iterative process. Once an application has been developed and adopted, software developers are reluctant to rewrite it. Given the time and effort invested in establishing a successful code base, it is far preferable to leverage the equity of an existing application rather than start from scratch. The broad and rapid adoption of the OOD (Object Oriented Design) approach to software development is evidence of this preference.

The complexity and extensive coding traditionally associated with protocols for parallel distributed processing have fostered the view that the task of modifying code for Grid Computing is something of a black art. Developers who understand the algorithm(s) of their software application typically lack the highly specific knowledge necessary to understand and effectively implement the available protocols, and find it impractical to invest the amount of time required to gain the necessary expertise. Furthermore, it is often necessary to engage in further development to manage the grid infrastructure beyond the scope of the application.

ON THE CD The CD-ROM accompanying this book includes GridIron XLR8™, a new application development tool that simplifies the process of implementing application-embedded parallel processing. GridIron XLR8 consists of two parts: An application developers' toolkit, or SDK, comprised of APIs that are added to the source code of a computationally intensive application, documentation, sample applications, and other tools and materials to assist a software developer in modifying their code; and runtime software that is installed on each computer in a grid providing the processing power.

GridIron XLR8 provides APIs at a high level of abstraction, so that developers do not have to worry about communications level programming. Instead, they simply work with familiar variables and data as they would with a serial program.

The GridIron XLR8 runtime software allows the computers on the grid to discover each other and automatically set up a processing network, distribute work, and recover from failure. This autonomic capability eliminates the need to code or manage the processing environment outside the application.

ON THE CD GridIron XLR8 is used in the subsequent example describing the process of grid-enabling a mainstream software application. Versions 1.2 of GridIron XLR8 for Windows and Mac OS X are provided on the CD-ROM and may be freely used for development and testing purposes under terms of the enclosed license (reference location of license hard copy in book) The most recent version of GridIron XLR8 and a version supporting Linux are available for download at the GridIron Software Web site: *http://www.gridironsoftware.com.*

12.9 THE PROCESS OF GRID-ENABLING SOFTWARE APPLICATIONS

The process of grid-enabling a software application is fairly straightforward. Using the GridIron XLR8 application development tool, an experienced developer familiar with the software application to be grid-enabled should complete the code modifications in a reasonably short period of time.

A distributable algorithm or job can be equivalently expressed as one or more steps. The most time-consuming, processing intensive steps that will be distributed for grid processing must have the following three characteristics:

- They can be split into smaller tasks.
- Each task can be processed on a separate computer.
- The results from each task can be returned re-assembled into one final result.

12.9.1 Analysis

To grid-enable an application for distribution, it must first be analyzed prior to modification of the software source code.

12.9.1.1 Identifying Hot Spots

The first step in the code analysis is to identify the partition points at which the application is most appropriately split into smaller tasks. This is achieved by locating the computational *hot spots* in the application algorithm where a majority of the execution time is spent running an encapsulated sub-algorithm.

A hot spot region should be easy for a software developer familiar with the algorithm or the application domain to identify. They are typically found within portions of iterative code, i.e., nested FOR or WHILE loops. Execution profiling can be used to empirically identify partition points by highlighting which few lines of code account for a high percentage of the application execution time.

12.9.1.2 Tightly Coupled versus Embarrassingly Parallel Algorithms

Application algorithms fall into two different categories. Some algorithms are easily segmented into tasks that can be processed entirely independently. Given the obvious nature of their partition points these algorithms are said to be *embarrassingly parallel*. Scriptable applications that can be deployed using the batch queue invariably contain embarrassingly parallel algorithms; however, there are also many non-scriptable applications that contain embarrassingly parallel algorithms and which can be easily modified for distributed processing on a compute grid using the programmatic method. The MPEG encoder application referenced later in this chapter is an example of the latter.

Conversely, *tightly coupled* algorithms have dependencies on interim communication and exchange of results in the computational process. Tightly coupled problems do not lend themselves to batch queue distribution and require code modification in order to take advantage of parallel processing. Applications with tightly coupled algorithms generally require more effort and are more difficult to modify for parallel distributed processing through programmatic means.

As part of the partition point portion of the analysis, identification of the category of algorithm will allow for accommodations to be made for interdependent task processing and interim result sharing for those that are tightly coupled in nature.

12.9.1.3 Operating Environment

Once the partition points surrounding the application's hot spot have been identified, the second step of the analysis is to determine what specific items are required for each task to be processed on a separate computer.

This entails identifying the application requirements for local files, libraries, or databases, and special licensing or hardware requirements. These elements of the application's normal operating environment will need to be dynamically provided to the grid as part of the task if they are not statically pre-installed on all compute nodes. The GridIron XLR8 framework provides support for the automated distribution of these files.

12.9.1.4 Results Generation

The final step of the analysis is to identify what kind of results the task computation will generate and determine how task computation results are to be stored and/or processed. The developer should note the likelihood that task results in a distributed environment will arrive asynchronously in a different order than they were defined.

12.9.2 Application Modifications

Once an application has been carefully and thoroughly analyzed, the application code can then be modified using the GridIron XLR8 development tool to allow the distributed tasks to be split and sent to the grid for computation, and to re-assemble the individually returned task results into a single aggregate job result.

12.9.2.1 Defining Tasks

First, an instance of the application called the *distributor* needs to be created, based on code *outside* the computational hot spot identified during analysis. The distributor contains the GridIron XLR8 *defineTask* method, which divides the job into smaller tasks for processing on a separate computer (effectively setting up the hot spot computation). The GridIron XLR8 framework provides facilities to automatically associate and transmit any files required for each task to the grid compute nodes.

At run time, the GridIron XLR8 framework will repeatedly invoke the function to create new tasks whenever compute resources are available. Through *defineTask*, the distributor controls the size of each task. In practice, this is based on a modification of the control code (i.e., FOR or WHILE LOOP conditions) associated with the hot spot identified during the analysis phase.

Note that the task size can be varied based on expected constraints of the deployment environment, such as interconnect speed, whether the network is static or dynamic, reliability and availability of the network and computers, and the minimum and maximum number of computers making up the grid. Based on these constraints, a default task size is selected which maximizes the ratio of the task computation time relative to the time required for setup, result recompiling and communication. Ideally, the size of a task should be configurable so that optimization can be easily achieved without making further changes to the application code.

GridIron XLR8 framework will invoke *defineTask* until the entire job is complete.

12.9.2.2 Task Computation

The second step in code modification is to create an instance of the application called the *executor*, which defines a new function completely encapsulating the algorithm contained *inside* the selected hot spot. This is accomplished with the GridIron XLR8 *doTask* method.

This is where the computationally intense part of the application code goes. GridIron XLR8 distributes the *doTask* method to the compute nodes for execution. Upon completion of the computation, the framework will automatically return results back to the distributor.

For tightly coupled algorithms, it will be necessary for interim results to be communicated between tasks before the individual task results are returned to the distributor.

12.9.2.3 Result Re-assembly

The third and final step in code modification is to modify the distributor to define another new function containing the results-handling portion of the algorithm. This is the GridIron XLR8 *checkTaskResults* method, and it contains code originally found within or immediately following the application hot spot.

At run time, after processing is complete, the GridIron XLR8 framework will ensure all files and/or data defined by the executor have been transmitted back to the distributor.

Since it is likely task results will be returned in a different order than that in which they were defined, results may have to be stored until all results have been received, whereupon the final, aggregated result can then be generated.

12.10 GRID-ENABLING A MAINSTREAM SOFTWARE APPLICATION: AN EXAMPLE

It has been previously noted that no broad adoption of Grid Computing will occur unless there is sufficient demand driven by a quantity of useful, grid-enabled applications that can readily and easily take advantage of the grid. It has also been noted that there exist many software applications that are computationally intensive but are not considered to be high performance computing applications.

One such category of software is applications used for digital content creation. Many processes in the creation of digital audio, video, and graphics are computationally intensive. For example, rendering, compositing, animation and encoding/decoding either require large amounts of processing or handle large amounts of data, and in many cases, both.

Video encoding is one such computationally intensive digital content creation software process. Furthermore, there is evidence to suggest that the already computationally intense nature of video encoding will further increase in the near term.

Although effective hardware solutions exist to reduce the time required to encode digital video, they have certain shortcomings that considerably limit their practical use, particularly by less sophisticated mainstream technology users.

Grid Computing provides powerful processing capacity at low cost using commodity computing hardware. As such, it offers tremendous potential as a means of accelerating digital video encoding.

A number of academic studies have investigated the feasibility of distributed computing for the purpose of video encoding, and some have had demonstrable success in controlled laboratory environments. However, none have proven sufficiently robust for commercial use in harsher real world conditions.

Excerpted from the white paper "Distributed Video Encoding for Use In Commercial Software Applications,"[2] this example demonstrates the practical grid-enabling of a mainstream software application. It describes how an open source

---

[2] "Distributed Video Encoding for Use In Commercial Software Applications." Available at *http://www.gridironsoftware.com*

MPEG-4 software encoding application was grid-enabled to allow video encoding to be performed more rapidly on a compute grid.

12.10.1 Video Encoding

Video encoding is a process for compressing raw digital video data by several factors in order to make storage and transmission more practical.

The Moving Picture Experts Group (MPEG) is a working group of ISO/IEC (International Organization for Standardization/International Electrotechnical Commission) in charge of the development of international standards for coded representation of digital audio and video. Established in 1988, the group has produced MPEG-1—the standard on which such products as Video CD and MP3 are based, MPEG-2—the standard on which such products as Digital Television set top boxes and DVD are based, MPEG-4—the standard for multimedia for the fixed and mobile Web, and MPEG-7—the standard for description and search of audio and visual content.

MPEG-4 is the most recent of the three most commonly implemented MPEG standards for encoded digital video. MPEG-4 builds on the proven success of earlier MPEG-1 and MPEG-2 work in the fields of digital television, interactive graphics applications, and interactive multimedia. It provides the standardized technological elements enabling the integration of the production, distribution, and content access paradigms of the three fields. Successful distribution of encoding to the MPEG-4 standard thus demonstrates the broad applicability of distributed computing to video encoding for a wide range of uses.

MPEG-4 encoding involves a number of tasks, including: image processing; format conversion; quantization and inverse quantization; discrete cosine transform (DCT); inverse DCT; motion compensation; and motion estimation. These functions reduce the sequence of video data to a much smaller bit stream by reducing redundancies in space and time from image to image. Of these tasks, motion-compensated prediction is highly computationally intensive, and can require as many as several gigaflops per second.

12.10.2 The Need for Speed

There are three pronounced trends that suggest the already computationally intense nature of video encoding will further increase in the near term.

The growing popularity of HDTV (high definition television) suggests the high definition standard will soon eclipse the current National Television System Committee (NTSC) broadcast standard. Raw video to support the NTSC standard at a resolution of 720 by 480 pixels and a frame rate of 30 frames per second is delivered at a bit rate of 249 Mbps and requires approximately 1.9 GB of storage per minute. By comparison, the HD standard of 1920 by 1080 pixels at the same 30 fps frame rate is delivered at a bit rate of 1.5 Gbps and requires approximately 1.1 TB of storage per minute of raw video. With the bit rate increased by a factor of six and the storage requirement increased by over two orders of magnitude, HD has much more rigorous compression requirements than the current NTSC standard.

Secondly, in addition to new video *presentation* standards, each new *digital coding* standard is similarly satisfied only through greater computing power. The compute expense necessary to achieve software-based encoding has increased as the MPEG standard has evolved from MPEG-1 to MPEG-2 to MPEG-4. This trend suggests successive generations of the MPEG standard, or an alternate standard, will require more powerful processing to encode to the new specifications.

Finally, digital video (DV) cameras are among the fastest selling consumer electronics products. The popularity and proliferation of digital video cameras is in turn driving a demand in the consumer segment for digital video manipulation products, such as editors, compositing tools, DVD creation tools and disc burners, etc. Many of these tools output video in a number of different formats, and thus feature native video encoding capabilities. Previously available only to video professionals, more and more of these manipulation products are being introduced downstream into consumer markets. Less sophisticated mainstream users now expect the same powerful capability of these products at lower prices. To meet the demands of this market, the computational intensity associated with encoding must be accommodated with a solution that is simple to use and deploy, and must run on inexpensive, consumer-grade (commodity) computing hardware.

12.10.3 Current Solutions

Two general categories of solutions currently exist to accommodate the intense computing needs of MPEG-4 video encoding.

Various hardware solutions, such as hardware accelerator blocks or loosely coupled coprocessors, are available to speed the encoding process. Although hardware solutions are effective in reducing the time required to encode video to MPEG-4 standards, they have certain shortcomings. Most hardware solutions are single purpose. In addition to being limited in applicability to only video processing applications, in many cases they are further limited to specific codecs or encoding algorithms. Hardware solutions are also functionality restricted relative to software solutions and are less flexibly upgraded to accommodate new encoding technologies or changes in hardware environments, such as algorithm improvements or the introduction of new processors.

Software encoding solutions, on the other hand, afford superior portability and flexibility. However, the chief drawback to these solutions is the amount of compute time and processing power it takes for the software encoder to run. To date, the means of speeding software encoding is to provide more powerful CPU, often accomplished with SMP computers. This makes it both expensive and difficult to scale to achieve fast encoding.

The ability to distribute video encoding potentially offers the best of both worlds: The flexibility and functionality of software encoding inexpensively implemented and easily scaled using multiple low-cost computers.

12.10.4 Grid Deployment of Video Encoding

To date, a number of academic studies have investigated the applicability of distributed processing and data distribution methods for the purposes of video encoding. Some of these studies have substantiated the theoretical applicability of parallel and distributed computing to the encoding process, and some have shown successful implementation of certain key techniques.

The logical evolution of this preliminary academic work is to develop a complete implementation that can allow distributed video encoding to be successfully migrated from the controlled and delicate environment of the research lab to harsher commercial environments.

12.10.5 Requirements for Broad Marketplace Adoption

To be viable for use in widely circulated commercial video encoding applications (encoders, codecs, compression algorithms, etc.), distributed computing must not only be *technically feasible*, but *practically implemented*. Although the feasibility of certain aspects of distributed video encoding have been proven in the lab, as yet no complete, end-to-end, working solution has been demonstrated in a heterogeneous, multiple computer (grid or cluster) environment. Furthermore, although it may be *feasible* to successfully apply distributed computing methods to video encoding, there are certain *practical* criteria that must also be met to ensure commercial viability. In particular, the implementation of distributed computing for video encoding must be:

- Simple enough for application end users to deploy and use without having to acquire special skills or knowledge
- Sufficiently robust to operate reliably outside of controlled lab conditions where environments may have varied hardware, operating system, and network elements
- Fast and easy for the commercial software developers to integrate into their video encoding applications

12.10.6 Overview of MPEG 4 Encoder

The MPEG4IP project provides an MPEG and IETF standards-based system for encoding, streaming, and playing MPEG-4 encoded audio and video.

MPEG4IP makes available an end-to-end system to explore MPEG-4 multimedia. The package includes many existing open source packages and the "glue" to integrate them. It is a tool for streaming video and audio that is standards-oriented and free from proprietary protocols and extensions.

The MPEG4IP package includes an MPEG-4 AAC audio encoder, an MP3 encoder, two MPEG-4 video encoders, an MP4 file creator and hinter, an IETF standards-based streaming server, and an MPEG-4 player that can both stream and playback from local file.

MPEG4IP's tools are available on the Linux platform, and most of the various components have been ported to Windows, Solaris, FreeBSD, BSD/OS, and Mac OS X.

The MPEG4IP MPEG-4 encoder is a software-based encoder in C++ with a relatively well-structured source code environment. The open-source, multi-platform, standards-based nature of the encoder makes it ideal as a broadly representative example of a video encoder that can be modified to take advantage of distributed computing. It should be noted, however, that the MPEG4IP MPEG-4 encoder is not enhanced or optimized to the degree found in most commercial MPEG-4 encoders and codecs. In most cases, such commercial software exhibits far superior performance, some by as much as an order of magnitude or more. For this reason, it is important to caution that the *relative* performance between the non-distributed and distributed versions of the *same* encoder be assessed, rather than directly comparing encoding time between the MPEG4IP encoder (distributed or not) with that of other encoders.

12.10.7 Overview of GridIron XLR8

GridIron XLR8 is a product that allows software developers to add the speed of distributed computing to commercial software applications. XLR8 enables computationally intensive software applications to run faster on multiple computers.

GridIron XLR8 consists of two parts: An application developers' toolkit, or SDK, comprised of APIs that are added to the source code of a computationally intensive application, plus documentation, sample applications, and other tools and materials to assist software developers in modifying their code for processing by multiple computers; and runtime software that is installed on each computer in a network, providing additional processing power.

Using GridIron XLR8 to add distributed computing to the MPEG4IP video encoder is beneficial in three ways:

■ It provides the software developer unfamiliar with distributed computing techniques and message passing with a simple and rapid development environment.

■ It eliminates the need to code or manage the processing environment outside the application.

■ As an application-embedded solution, it is simple for end-users to work with the final compiled and installed version of the distributed video encoding software.

GridIron XLR8 reduces the complexity of embedding distributed computing within an application by providing all the necessary programmatic elements at a high level of abstraction. All of the job control logic can be defined and controlled through the use of just six XLR8 job control functions and four job execution methods provided by application plug-ins. Additional XLR8 functions are available for administration, management, and data marshalling. By comparison, protocols such as MPI are significantly more complex, with some 380 primary calls.

The pre-built GridIron XLR8 runtime software entirely eliminates the need to code or manage the processing environment outside the application. The intelligent peer-to-peer architecture of the GridIron XLR8 runtime software allows computers to discover each other and automatically set up a processing network, distribute work, and recover from failure. The high-level programmatic abstraction combined with the intelligent runtime software significantly reduces the development effort currently associated with distributed computing. Furthermore, this autonomic capability eliminates the requirement for an administratively intensive middleware layer, making the solution simple to deploy and manage.

Finally, GridIron XLR8 is embedded directly into the software applications. Once compiled and installed, users can benefit from the speed of distributed computing without having to change the way they use the application and without learning special skills.

GridIron XLR8 version 1.1 is included with the CD-ROM accompanying this
ON THE CD book and is available for download at *http://www.gridironsoftware.com*.

12.10.8 Distributed Computing Strategy

Key to the successful distribution of the encoding application is to follow the process of grid-enabling software applications described above to determine how to successfully divide video data into segments that can be processed in parallel on multiple computers.

The MPEG-4 specification makes use of a hierarchy of video objects to represent content. A *Video Session* is the top tier in the hierarchy, representing the whole MPEG-4 scene. Each *Video Session*, or scene, is populated with *Video Objects*, which can be encoded as single or multiple *Video Object Layers*. A *Video Object Layer* is in turn comprised of a *Group Of Pictures (GOP)*.

A Group Of Pictures is a collection of three different picture types that typically occur in a repeating sequence:

- *I-pictures*, or intra pictures, are pictures that are moderately compressed and coded without reference to other pictures. These are used as a reference points in the decoding process.
- *P-pictures*, or "predictive" pictures, take advantage of motion-compensated prediction from a preceding I- or P-picture to allow much greater compression.
- *B-pictures*, or bi-directionally-predictive pictures use motion-compensated prediction from both past and future pictures to allow the highest degree of compensation.

GOPs have the same number of pictures per sequence and are similar in size. Segmenting the video data for distribution at the GOP tier in the hierarchy takes advantage of these characteristics to effectively treat the MPEG-4 encoding process as an embarrassingly parallel software algorithm. The size and structure of the GOPs makes them convenient to distribute for parallel processing at a reasonable level of granularity, to accommodate typical CPU and bandwidth capabilities, and to achieve reasonable load balancing across multiple processors.

12.10.9 Implementation

This section discusses the implementation process undertaken to grid-enable and test the application.

12.10.9.1 Application Modification

The first and most time-consuming step of the implementation was the identification of the appropriate partition points in the encoder, i.e., areas where data are encoded into I-, P-, and B-pictures.

Of the 1,089 source files that make up the MPEG4IP suite, 161 were identified as tools relative to encoding. Given the high level of abstraction of the GridIron XLR8 APIs and the self-contained nature of the GridIron XLR8 runtime software, it was necessary to modify just three of these files in order to implement distributed computation: The Encoder.cpp, Sesenc.cpp, and Sesenc.hpp files.

Files were modified as required in order to segment and distribute the video data based on the GOP strategy. The number of frames in a GOP is configurable, and based on experimentation to allow for optimization with the given hardware environment and video data, a high level partitioning was implemented in segments of 20 frames. In other words, each sequence of 20 frames was essentially treated as a small, independent movie.

The raw video data were comprised of a total of 1,824 frames. At a partitioning of 20 frames, this yielded a total of 92 partitions. Each of the partitions was treated as an individual processing task, and encoded by an individual XLR8 runtime software peer. This yielded 92 compressed files, which were asynchronously returned and appropriately multiplexed back into a single, new compressed file. Encoding occurred at a target bit rate of 1.5 Mb per second.

12.10.9.2 Hardware

A grid comprised of 13 IBM xSeries 335 servers was used for this implementation, each with dual 2.0 GHz Intel XEON processors, 1.0 Gb RAM, and the Windows 2000® Server operating system. The modified MPEG4IP MPEG-4 software encoder was installed on one of the 13 machines. Although fully capable of acting as a processing node, this server was only used to launch and run the encoder application. Only the GridIron XLR8 peer runtime software was installed on the remaining 12 xSeries computers, which were configured as individual servers and connected together with a gigabit Ethernet switch.

12.10.9.3 Data

The raw video data used for encoding were (approximately) 60-second NTSC broadcast quality (YUV [4:2:0]) at a resolution of 720×480 pixels and with a frame rate of 29.97 fps. The size of the uncompressed file was 923,400 KB.

12.10.9.4 Hyperthreading

There are a number of currently available technologies that provide the facility for performance improvement through coprocessor and software optimizations, such as vectorization (e.g., AltiVec), Single Instruction Multiple Data (SIMD), Pthreads, SSE2, etc. One such technology is hyperthreading.

Hyperthreading is an evolving Intel processor technology (first available on Intel's XEON server processors and now being delivered on all desktop 3.06 GHz+ processors) that provides dual simultaneous execution of two threads on the same physical processor. Performance improvements for most multi-threaded applications range from a typical 5 percent to a current theoretical maximum of approximately 30 percent.

Hyperthreading was utilized in this implementation to demonstrate that such technologies are complimentary to distributed computing and will achieve cumulative performance improvements.

Given the support for hyperthreading by both the IBM xSeries 335 hardware and the Windows 2000 Server operating system used for this implementation, it was possible to deploy four instances of the GridIron XLR8 runtime software per node (machine), i.e., one independent XLR8 computing peer per each of the two logical processors provided for each single physical processor as a function of the hyperthreading technology. This provided each dual processor node with the ability to execute (as many as) four independent parallel processing tasks.

12.10.9.5 GridIron XLR8 Runtime Software

The pre-built GridIron XLR8 runtime software manages the processing environment outside the application and allows computers to discover each other and automatically set up a processing network, distribute work, and recover from failure.

As mentioned previously, the hyperthreading capability of the hardware and operating system allowed four instances of the Windows version of the GridIron XLR8 runtime peer to be installed per machine. There were thus a total of 48 peers deployed on the 12 servers acting as processing nodes in this implementation. Each peer has an installed footprint of 24 MB.

12.10.10 Results

This section discusses the performance improvements seen by grid-enabling the MPEG4 application.

12.10.10.1 Output

The implementation described above generated a compressed and encoded MPEG-4 file that was decodable and playable at expected levels of quality using an existing MPEG player.

12.10.10.2 Compression

The resultant MPEG-4 compressed file was 13,798 KB, representing a reduction of about 98.5 percent from the original raw video data.

12.10.10.3 Speed Improvement

Table 1 shows the results of the implementation with multiple compute nodes:

A graphical representation of the results showing the speed-up factor by the number of (physical) processors (see Figure 12.1.) shows that essentially linear speed-up was achieved by the implementation for up to 12 CPUs, i.e. the encoding ran twice as fast with two physical processors, four times as fast with four processors, eight times as fast with eight processors and twelve times as fast with twelve processors. With 16 processors or more, the implementation showed near linear speed-up, with the speed-up factor beginning to trend away from linear performance.

TABLE 12.1 Encoding results by nodes, physical and logical CPUs, and number of peers

Nodes	Physical CPU	Logical CPU	No. of XLR8 Peers	Projected linear time (seconds)	Actual time (seconds)	% Deviation
1	1	2	2	8,820	8,820	0.00%
2	4	8	8	2,205	2,004	−10.03%
4	8	16	16	1,103	1,053	−4.70%
6	12	24	24	735	739	0.54%
8	16	32	32	551	615	10.37%
10	20	40	40	441	523	15.68%
12	24	48	48	367	472	22.14%

12.10.10.4 Development Effort

The bulk of the work in implementing a distributed encoding solution using the MPEG4IP encoder and GridIron XLR8 was in identifying the appropriate partition points in the encoder, and testing, rather than code work. The breakdown of effort associated with the described implementation was approximately one person-week of analysis, one week of code modification, and one week of testing and benchmarking.

It was necessary to modify just 3 of the 1,089 source files included with the MPEG4IP suite. In total, approximately 100 lines of code were modified and 1,000 lines of supplemental code, including comments, were added.

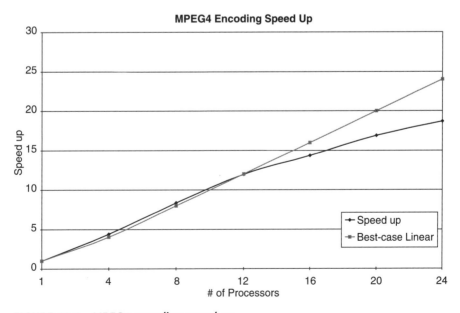

FIGURE 12.1 MPEG4 encoding speed up.

12.10.10.5 Impact of Hyperthreading on Results

After some preliminary system tuning of the application, an improvement over the non-hyperthreaded implementation of up to 12.6 percent was measured. This performance improvement appears to have been achieved due to two factors: The explicit benefits of hyperthreading, and the added side effect of job queuing.

12.10.11 Next Steps

An important next step is the execution of this implementation on 12 + 1 nodes with 24 physical and 48 logical processors (through hyper-threading) dedicated to parallel processing displayed linear improvement with up to 12 CPUs, and near linear improvement with 16 to 24 CPUs.

Even though this configuration is probably typical of a compute cluster that might reasonably be implemented for video encoding, further testing in a larger environment is required to measure scalability and identify thresholds for performance degradation on larger numbers of compute nodes.

Further work on partitioning is required to fully optimize performance. Early experimentation indicates that optimal partitioning by number of frames is relative to the number and type of CPUs implemented. For example, reduction of the number of frame rates used in a partition has resulted in faster processing on larger numbers of CPUs, thereby achieving closer to linear speed up beyond 16 CPUs. It is expected that dynamic determination of the optimal number of frame rates base on the data, the encoder, and the number and type of available processors will yield further linearity and scalability with more than 12 compute nodes.

The MPEG4IP MPEG-4 encoder used in this implementation is a single-pass, CBR (constant bit rate) encoder. Current state-of-the-art video encoders implement a multi-pass, VBR (variable bit rate) approach for encoding, which yields a higher quality result. The implementation should be replicated in such an encoder to validate the hypothesis that a similar linear improvement in speed-up can be achieved.

Generally, this implementation represents a first effort at distributed video encoding for use in commercial software applications. Developers must to continue to enhance the application test bed as they strive for excellence in areas of task queuing, hyperthreading, and scheduling policies. It is expected that these additional refinements will result in further improvements in performance.

12.10.12 Grid-Enabling Video Encoding Summary

The described MPEG-4 software encoding solution represents both a feasible implementation of distributed computing for the purposes of video encoding and meets key practical requirements essential for commercial viability.

The implementation successfully achieved approximately linear speed-up of MPEG-4 encoding through distributed parallel processing across a grid comprised of up to 12 compute nodes. The result was a reduction of encoding time by a factor of 20, from 2 hours 33 minutes on a single machine to 7 minutes 4 seconds. With greater knowledge of encoding and the specific MPEG-4 encoder application, even further performance improvements can be achieved through additional optimization. Some limited experimental tuning to date has achieved a reduction by an even greater factor, to 6 minutes, 0 seconds.

This implementation also demonstrated that technologies that provide the facility for performance improvement through coprocessor and software optimizations can be used in conjunction with distributed computing to achieve cumulative performance improvement. In this case, hyperthreading was used in addition to parallel distributed processing to achieve an additional performance increase of up to 12.6 percent.

Furthermore, this implementation satisfies key requirements essential for commercial viability and deployment in harsh, real-world conditions. The GridIron XLR8 technology is application-embedded and provides a self-configuring runtime environment that makes the solution simple for application end users to deploy and use without having to acquire special skills or knowledge. The self-configuring, zero administration of the runtime environment also makes the demonstrated implementation sufficiently robust to operate reliably outside of controlled lab conditions.

The high-level programmatic abstraction of the GridIron XLR8 APIs combined with the intelligent runtime software significantly reduced the development effort traditionally associated with distributed computing. The implementation described was achieved with three weeks effort by a single individual, and required minimal code modification to a small number of source files relative to the entire application. In fact, it is likely this represents a conservative to high estimation of effort given that the implementer was not a domain expert in encoding. It is expected that the analysis and coding would take place over a much shorter period time when performed by developers already familiar and well acquainted with the encoding algorithm and code base.

Although a specific MPEG-4 encoder was used, the implementation described should be applicable to other MPEG-4 encoders and codecs, to other MPEG encoding processes, and to other methods of video encoding in general. The MPEG-1 and MPEG-2 standards utilize the same basic techniques of motion compensation and transform coding addressed by this implementation. It is also possible this technique for distributed encoding can be applied to emerging new technologies that require substantial compression, such as JPEG2000.

12.11 CONCLUSION

Grid Computing has reached a critical juncture. The technology is at a watershed moment where success or failure now rests solely on its ability to deliver on its much-hyped promise.

To date, attention has almost exclusively focused on the supply-side aspects of utility computing. Most effort has been invested in developing grid infrastructure technologies.

It will be the marketplace, however, that determines the ultimate demand for grids and the viability of the utility computing model. End users drive this demand, and the applications they use are their touch point with the grid. Attention must therefore shift rapidly to their practical needs and how this must be accommodated in the applications they use.

The demand-side stakeholders are application end users, business enterprises that represent a special class of aggregated application users, and software application developers. Although Grid Computing can potentially offer great benefit in the form of powerful processing, increased application speed, and resource utilization, it will only be adopted if these benefits are not outweighed by the inconvenience or difficulty incurred in order to use it. For this reason, it is vital that grid-enabled applications are simple to use, simple to manage, and simple to develop.

To date, the use of Grid Computing has been mostly limited to high performance computing and large enterprises. Only those with mission-critical or highly strategic computing requirements have found it worthwhile to undertake the onerous task of modifying non-scriptable applications for distributed processing, and DRM solutions for batch queue distribution of scriptable applications are appropriate for the many-to-many mapping of users to resources typically found only in the large enterprise.

The significant potential benefits that Grid Computing promises can be delivered to a dramatically larger universe by grid-enabling computationally intensive applications that are not necessarily high performance applications, broadening the use of scriptable applications beyond many-to-many environments, and grid-enabling non-scriptable applications.

ON THE CD Given the complexity and extensive coding traditionally associated with protocols for parallel distributed processing, few applications have been grid-enabled to date; however, a new application development tool included on the CD-ROM accompanying this book simplifies the process of implementing application-embedded parallel processing.

Using the GridIron XLR8 application development tool, an experienced developer familiar with a software application to be grid-enabled should complete the code modifications in a reasonably short period of time following the straightforward process of analysis and code modification described.

The grid-enabled MPEG-4 software encoder demonstrates that it is simple for a developer to modify the source code of a non-scriptable, computationally intensive application that delivers linear speed up on a grid with as few as two computers. Furthermore, once the code is modified, it is simple to use and manage, requiring only the installation of the GridIron XLR8 runtime software on the computers comprising the grid. No other configuration or management is necessary, and there is no change to the way in which the MPEG-4 encoder is used.

Making software applications grid-ready is not merely possible and practical; it is *vital* for the grid to enter the fabric of mainstream computing. Without a sufficient number of grid-enabled applications, Grid Computing will be relegated to the graveyard of novel technologies that failed to serve a useful purpose.

13 Application Integration

Jikku Venkat

United Devices

In This Chapter

- Classifying Applications
- Integrating Applications to Grid Middleware

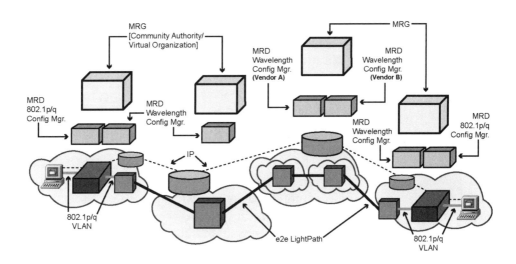

13.1 INTRODUCTION

The grid has been defined by Foster et al., as the "Software and technology infrastructure to support the coordination and sharing of resources in dynamic, distributed virtual organizations."[1] There are several implications of such an infrastructure that have an impact on the motivation and types of applications that are suitable for execution on a grid as opposed to more conventional forms of computing, where the resources are dedicated and static.

- Coordination and sharing of resources—Because resources are being shared, applications may not be intrusive to other users of the resource. Applications need to explicitly specify their resource requirements to ensure coordination.
- Dynamic and distributed virtual organizations.
  - Wide range of resource types—heterogeneous.
  - Resources are dynamic, i.e., they are not always available.
  - New resources are constantly added.
  - Distributed resources; no shared state and a (sometimes large) cost associated with communication between distributed nodes.
  - A virtual organization implies the grid infrastructure must provide the appropriate level of security and authentication to enable the usage of these distributed resources by authorized entities across the grid.

There are several motivating factors for deciding to use the grid as an infrastructure to run applications. Probably the most important motivation is to be able to use the vast amount of compute and storage resources across the Internet in a coordinated manner to get previously unattainable performance for certain classes of applications.

With the advent of storage and file systems that can be shared across a network, it has become feasible for application users to create large databases of information that could then be shared by their colleagues around the world, using grid technology.

Often, the motivation in selecting the grid as a platform for running applications is the availability of more capacity than what is available within the domain of the application or its user. Even during peak load times within a particular domain, an application user would be able to submit his job locally into the grid and have the job run on a remote set of resources that are available immediately on the grid rather than have to wait until similar resources become available locally.

Another benefit to running applications on the grid is the availability of a heterogeneous set of resources that make it possible for an application to run on the resource that best matches the requirements of the application.

---

[1] Foster, Kesselman, Nick, and Tuecke, *The Physiology of the Grid*, Open Infrastructure WG, June 2002.

The vast amount of computing and storage resources available on the grid make it possible to solve problems that were not computationally tractable on dedicated supercomputers and clusters.

Other use cases for the grid include some instances where the application is not necessarily compute-intensive. For example, the heterogeneity, geographical diversity, and environmental diversity of a vast number of resources on the grid can be used to test applications and servers. An instance of this is the use of the grid to load and stress-test Web servers. This involves a coordinated action from a vast number of resources to perform some real, deep transactions on a Web application to study its performance and functionality in the presence of a real load.

As grid infrastructures evolve, it will become more feasible to define workflows that utilize several existing application services across the grid to solve a larger problem. This use of grid is sometimes referred to as a *collaborative grid*.

13.2 APPLICATION CLASSIFICATION

The integration plan to enable an application to run on a grid is very dependent on the type of application being considered. In this section, applications are classified along a number of dimensions. These dimensions are relevant to enabling the application to run on a grid. The dimensions are parallelism, granularity, communications, and dependency. How an application is classified affects the way in which it can be migrated to a grid environment. These dimensions are not mutually exclusive—a specific application can be characterized along most, if not all, of these dimensions.

13.2.1 Parallelism

This discussion will focus on a well-known classification scheme attributed to Flynn.[2] This classification scheme is a useful one to consider because it has a significant impact on how the application is integrated to run on a grid.

13.2.1.1 Single Program, Single Data (SPSD)

These applications are simple sequential programs that take a single input set and generate a single output set. There are many motivations for running these applications in a grid environment.

First, a vast number of computing resources are immediately available. In this case, the grid is used as a throughput engine especially when many, many instances

---

[2] Flynn, M.J., "Some Computer Organizations and Their Effectiveness," *IEEE Trans. on Computers* 24 (9), pp. 948-960, Sept. 1972.

of this single program need to be executed. Leveraging resources outside the local domain can vastly improve the throughput of these jobs. An example of this type of application is the logic simulation of a microprocessor design. The design needs to be verified by running millions of test cases against the design. By leveraging grid resources, each test case can be run on a different remote machine, thereby improving the overall throughput of the simulation.

Second, remotely available shared data can be used. In this situation, the program can be made to execute where the data reside or the data can be accessed by the application via a "data grid." Thus, a grid environment enables controlled access to shared data.

It is a simple matter to integrate these applications to run on a grid, as no additional code development is required.

13.2.1.2 Single Program, Multiple Data (SPMD)

These are applications in which the input data can be partitioned and processed concurrently using the same program. This type of application comprises the majority of applications that utilize the grid today and covers a wide range of domains. Some examples are Finite Element Method evaluations using an MPI-based program and large-scale Internet applications such as *SETI@home*[3] and the UD-Cancer project that run on millions of PCs across the Internet.

The primary motivation for running this class of applications on a grid is to significantly improve performance and/or scope by scaling the application out to as many resources on the grid as possible. Both *SETI@home* and the UD-Cancer project are very good examples of the increase in scope.

13.2.1.3 Multiple Program Multiple Data (MIMD)

This is the broadest category of parallel/distributed applications. Here both the program and the data can be partitioned and processed concurrently.

The primary motivation for running this class of application is to improve performance. A secondary motivation is to ensure that resources are more optimally matched to application needs. For example, there might be partitions of an application that require a very tightly-coupled SMP system while another partition could be run in a more loosely-coupled environment or on a single machine.

13.2.1.4 Multiple Program Single Data (MPSD)

Multiple Program Single Data applications require different transformations to be applied to the same set of input data. In the grid environment, these transformations can be carried out concurrently. This category of applications is very rare.

---

[3] Seti@Home Internet Computing, *http://setiathome.ssl.berkeley.edu*

13.2.2 Communications

The communications cost associated with an application is characterized by the following:

- The cost of initial movement of data and programs to grid resources prior to the computation. This cost can be avoided on the grid in cases where the data or program is already available at the remote grid resource.
- The cost of communication while the application execution executes. This cost is a function of the frequency and size of communication.

The communication costs are obviously very dependent on the performance of the underlying grid network infrastructure.

13.2.3 Granularity

Applications can be classified based on the granularity of their programs as either coarse-grained or fine-grained. The granularity specifies how long a program can execute before it needs to communicate with other programs in the application. The granularity coupled with the cost of communication between the components of the application will dictate the amount of overhead associated with running the application remotely on the grid.

13.2.4 Dependency

A trivial case is when the programs within an application have no dependencies between them and each program can be scheduled and executed independently of the others. So long as the initial communications cost is reasonable compared to the granularity of the independent programs, the application would be a good candidate for deployment on the grid.

13.3 GRID REQUIREMENTS

This section outlines the key functional requirements of grids.

13.3.1 Interfaces

Users and applications have traditionally accessed grids using simple command-line tools and programming APIs. Command line tools such as 'qsub' and 'qstat'[4]

---

[4] PBS, *http://www.openpbs.org/main.html*

offer ways for users to interactively submit and query the status of their jobs. These tools can be further extended through *wrapper* scripts that hide the job creation and management complexities from the user. Similar to command-line tools and wrapper scripts, programming APIs provide a way for developers to embed job submission and management functions into an application-specific wrapper program. In either case, users execute an application-specific wrapper script or program and input the data required for the job.

The functionality provided by command-line tools and programming APIs is usually limited to running SPSD-type applications. These limitations have led to the development of more sophisticated interfaces that allow the creation of single- and multi-dimensional job arrays to support SPMD-, MPMD-, and MPSD-type applications. Some command-line tools now support job arrays with awkward syntaxes to specify multi-dimensional arrays and dependencies among jobs within an array or across arrays.

The advent of Web, XML, and Web Services has had a tremendous impact on grid interfaces. Web browsers are replacing command-line tools, enabling users to remotely access the grid from anywhere and at any time. The use of a Web Services description language (WSDL) to describe jobs, applications, and data in conjunction with Web Services provides a very powerful programmatic interface to the grid. Unlike traditional programming APIs, which require the developer to be bound to a specific language and toolkit, Web Services provides a language-independent interface for programmers to develop complex wrappers using their favorite programming language. Because Web Services is a well-defined standard, grids based on it will easily integrate into existing environments. The Open Grid Services Architecture (OGSA) is in the process of defining a set of Web Services Description Language (WSDL) interfaces for creating, managing, and securely accessing large computational grids.

13.3.2 Job Scheduling

The primary aim of a grid is to provide transparent and efficient access to remote and geographically distributed resources. A scheduling service is necessary to coordinate access to the different resources such as network, data, storage, software, and computing elements that are available on the grid. However, the heterogeneous and dynamic nature of the grid makes the task of scheduling jobs on such a distributed system fairly complicated. Schedulers need to transparently automate the three essential tasks required to schedule jobs in a grid environment, viz., resource discovery, system selection, and job execution.

At a very basic level, jobs, once submitted into the grid, are queued by the scheduler and dispatched to compute nodes as they become available. However, grid schedulers are typically required to dispatch jobs based on a well-defined algorithm such as First-Come-First-Serve (FCFS), Shortest-Job-First (SJF), Round-Robin (RR), or Least-Recently-Serviced (LRS). In addition, schedulers have to support advanced features such as:

- User requested job priority
- Allocation of resources to users based on percentages
- Concurrency limit on the numbers jobs a user is allowed to run
- User specifiable resource requirements
- Advanced reservation of resources
- Resource usage limits enforced by administrators

13.3.3 Data Management

Data processed by jobs are typically sourced at submission time or obtained by the compute node from a shared file system (NFS, AFS, DFS) at execution time. In the latter case, the location of the data is input at submission and each node is required to have access to the specified location. These mechanisms in their current form work well when the input data size is small or when all nodes in the grid have access to a global shared file system. However, neither solution is suitable for large grids, where nodes may be distributed across a wide-area network and have limited or no access to global file systems.

Typical data sizes being processed in both commercial and R&D are in gigabytes requiring very optimal data management methods to move and store data on wide area networks. Several hardware and software solutions are available with features such as distributed data caching and replication. These technologies, combined with a uniform namespace, can virtualize access to data and provide users and compute nodes a seamless way to store and retrieve data.

13.3.4 Remote Execution Environment

Jobs executing on a grid require the same environment as they would if executed on the job submitters' machine. These include environment variables, runtime environments such as Java, Small Talk, or .NET CLR, and libraries such as C language runtime and operating system libraries. This total environment may be already available on the compute node or may have to be created by the job on node prior to execution. Many grid frameworks provide tools and well-documented procedures to create such environments.

13.3.5 Security

Grid Computing covers a broad spectrum of interconnected computer systems, ranging from tightly managed supercomputers to more independently used desktops and laptops distributed over a wide area network. Despite the definition and observation of strict security policies, grids are still vulnerable to security attacks at all network points, with the greatest threat remaining at the desktop level. Some grid frameworks offer end-to-end security that includes *sand-boxing* of jobs during execution on compute nodes. The sand-boxing feature enforces a security wall between the job execution environment and the compute node on which it is running. For example, a desktop user will be unable to view or manipulate the grid job running on his computer.

The following are some basic features required to operate a secure grid:

- *Authentication* is used to positively verify the identity of users, devices, or other entity in the grid, often as a prerequisite to allowing access to resources in the grid. Authentication is accomplished using passwords and challenge-and-response protocols.
- *Confidentiality* is the assurance that job information is not disclosed to unauthorized persons, processes, or devices. Confidentiality is accomplished through access controls, and protection of data through encryption techniques.
- *Data Integrity* refers to a condition when job data are unchanged from their source and have not been accidentally or maliciously modified, altered, or destroyed. Data integrity is accomplished through checksum validation and digital signature schemes.
- *Non-repudiation* is a method by which the sender of job data, such as the grid scheduler, is provided with proof of delivery and the recipient, such as the compute node, is assured of the sender's identity, so that neither can later deny having processed the data. This method requires strong authentication and data integrity, as well as verifying that the sender's identity is connected to the data that are being submitted.

13.3.6 Gang Scheduling

The term *gang scheduling* refers to all of a program's threads of execution being grouped into a gang and concurrently scheduled on distinct processors. Furthermore, *time-slicing* is supported through the concurrent preemption and later rescheduling of the gang. These threads of execution are not necessarily POSIX threads, but components of a program which can execute simultaneously. The threads may span multiple computers and/or UNIX processes. Communications between threads may be performed through shared memory, message passing, and/or other means.

Each gang-scheduled job requires all its threads to be started and stopped concurrently, causing synchronization among threads to occur at the start and end of the job. In addition, if any specific thread aborts abruptly, all other threads must be terminated and the entire job restarted from the beginning. There are several proposals currently under consideration that enable fault-tolerant execution of gang scheduled jobs.

13.3.7 Checkpointing and Job Migration

Checkpointing enables a job to take a snapshot of its state, so that it can be restarted later. There are two main reasons for checkpointing a job: fault tolerance—where the job must recover from a failed compute node failure, and load balancing—where a job on a overloaded compute node must be migrated to a node with lesser load.

A long running job can be checkpointed periodically during its run. If the execution node fails for any reason, the job can resume execution from the last checkpoint when the node recovers rather than start from the beginning. The job can also be restarted on a different node if the original node is unavailable.

Sometimes one node is overloaded while the others are idle or lightly loaded. Jobs can checkpoint one or more jobs on the overloaded node and restart the jobs on other idle or lightly loaded nodes. Job migration can also be used to move intensive jobs to avoid interference with users or other programs running on the same node.

Checkpointing may be implemented at two levels: kernel level and user level. In the former case, the operating system transparently supports the checkpointing and restarting process without any changes to the application. In the latter case, the application can be coded in a way to checkpoint itself periodically. When restarted, the application looks for the checkpoint files and restores its state.

13.3.8 Management

Grids typically contain collections of geographically distributed workstations, servers, clusters, and supercomputers. Recently, applications and data have also been added to the grid, requiring very complex processes and tools to manage the entire system.

Grid management is a non-trivial task requiring the underlying framework to provide some basic but very important functions:

- *Virtualization* is the ability for administrators and job submitters to transparently access and manage compute resources, applications and data from anywhere in the grid. These entities, once added to the grid, may be transparently located anywhere and updated periodically without affecting the overall operation of the grid.
- *Provisioning* enables heterogeneous resources to be grouped based on their capability, location, or territorial control. This controlled sharing of resources delivers the benefits of Grid Computing while preserving existing decentralized IT policies. IT administrators are able to specify who can access each group and with what privileges and during what times.

■ *Accounting* provides IT managers the ability to assess the utilization of resources and re-provision them based on user demand and business priorities. In addition, users may be charged based on the type of resources and the duration for which they use them.

13.4 INTEGRATING APPLICATIONS WITH MIDDLEWARE PLATFORMS

The simplest classes of applications to integrate are SPSD. There are some basic mandatory functions that are required:

■ Prepare the application for remote execution.
■ Login to the grid or have the appropriate access credentials on the remote machine.
■ Submit the application and data for execution on the grid platform.
■ Retrieve results from the remote machine where the application executes.

Middleware products include support for one or more of these steps. In the simplest case, these applications simply run "as is." No source code modifications or additional software development is required. Middleware products provide varying levels of abstraction to the end user by performing one or more of the above steps. They abstract the job submitter from the remote machine and its execution environment. The user simply specifies the execution environment and the middleware identifies the appropriate resource, sets up the execution environment (code, data, environment), accesses the resource, and runs the application. For example, there are command-line utilities to submit jobs and retrieve results from the GridMP. The grid itself is completely transparent to the end user. On the other hand, packages such as the Globus Toolkit provide a set of services for each of these steps and the end user has to develop the accompanying software to integrate these steps. An alternative access method is via a browser-based console. The console provides a user or an administrator with an easy-to-use interface to install applications, register data sets, submit jobs, and retrieve results.

SPMD applications require additional functions to enable them to run on a grid. The simplest example of a SPMD application is a *parametric* simulation (a.k.a. Monte Carlo simulation). This case does not require any additional functions and simply represents a special instance of SPSD. In the case of true SPMD applications, the input data need to be partitioned and the partial outputs need to be merged to form a final result. An example is virtual screening, where a database of drug-like molecules is tested against one or more protein targets. The molecular database is partitioned, and each instance of execution works on this partitioned data. The partitioning of data, submission of individual pieces of work (called *workunits*), and retrieval and merging of results is accomplished by a separate piece of software, which we refer to as an application service. The application service can be accessed by a number of end users or clients simultaneously to run jobs.

Both SPSD and SPMD applications sometimes include a GUI that an end user interacts with. In these cases, the best approach for integration is to separate the GUI from the core application so that the GUI runs on the end-user's machine while the core application is run on a grid resource. In this case, the application service is integrated with the GUI and interacts with the GUI during job submission. An example of this is Accelrys' LigandFit application. This application included a separate graphical user interface that was the primary interface for end users. The grid application service was invoked from within this GUI application. End users continued to interface via the graphical front-end and the grid was completely transparent.

MPMD applications require modifications to the application source code so that the different program partitions are available as individual binaries. Typically, these programs have some dependencies that require interaction between the program partitions during execution. This can be done via a message-passing interface such as MPI or PVM. The final merge step is also more complex because there might be additional computation that needs to occur after all of the program partitions complete execution.

13.4.1 Application Preparation Example

A simple command-line utility, *buildmodule*, is available with the Grid MP. This utility packages the application executables along with metadata that describes the execution environment. The metadata is described within a module definition file. An example module definition file is shown below:

```
<?xml version="1.0" ?>
<module>
        <exe name="mytask.exe" ?>
        <packages>
        <package name="workunit" encoding="tar" ?>
        <package name="resdata" encoding="tar" />
        </packages>
        <cmdline value="%MYVALUE1 % output %MYVALUE2% %MYVALUE3%" />
        <output encoding="gzip" />

</module>
```

The metadata includes the name of the executable (mytask.exe), data packages that are used with the application (workunit and resdata), and the command line invocation for this particular task. These are specified as variables such as MY-VALUE1 that can be substituted during execution.

Once the application is packaged, jobs can be submitted against this application using different input datasets.

13.4.2 Issues in Application Integration

The goal of most grid middleware systems is to allow applications to be grid-enabled with minimal effort on the part of the application developer. Most grid middleware systems provide facilities that allow the developer to re-use existing components on the grid, and to use workflow primitives to compose larger applications using these existing primitives.

Legacy applications are typically applications that were not developed in-house and in most cases, the source code for these applications is not available for modification. Even in cases where the application can be modified, it is very expensive to develop and maintain a separate code base specifically for the grid. With this in mind, most grid systems support the execution of application code as-is, typically without requiring even a recompilation of the code.

An important step in enabling an application that cannot be avoided is the specification of scheduling policies, the parallelization of the application if desired, and basic job management functions that are typically integrated with the existing front end for the application. Using the existing front end for an application is very useful in maintaining the user experience while providing the performance and resource sharing benefits of the grid.

In the United Devices Grid MP platform,[4] for example, application codes can be run as-is on grid nodes. The additional development effort is required to write application scripts that take the input data and parameters, and issue the appropriate job submission calls to the grid, monitor the progress of the job, and retrieve results once the job has completed.

There are several other factors that affect the integration of applications on the grid. Due to the wide range of heterogeneous computing and storage resources available across a typical grid, it is important to virtualize access to these resources and provide mechanisms to allow applications to operate seamlessly across these resources, where possible. In some grid systems, applications are constrained to run on specific types of compute resources, while others like the United Devices Grid MP allow the registration of multiple executables for a single application, where each executable is compiled to run on a different platform. The Grid MP scheduler can now schedule any jobs running against this application to any available resources for which an executable has been registered.

---

[4] "Application Developers Guide," United Devices, *http://www.ud.com*

The overall performance of an application is best when the resource requirements of the application match the resources available on the machines on which it runs. Grid platforms typically approach this aspect of scheduling in several ways. Most traditional batch queueing systems such as LSF[5] and PBS[6] leave the matching to the user, and require that the job submitter select a queue for the job, where the queue is typically serviced by a specific type of machine. Other systems such as Condor[7] and Grid MP provide the user with mechanisms to specify the resource requirements for his job so that the scheduler can attempt to schedule an application to a machine that is a good match. A third approach is to use historical data for the execution of an application when making future scheduling decisions.

Because grid systems are typically spread across a wide area, the management of data and programs is critical for good performance of an application. Data management systems such as SRB[8] and Chimera[9] provide mechanisms to manage multiple copies of a dataset across the grid and to use this cached data to get improved performance for their applications. Another optimization is to use affinity scheduling, where the scheduler attempts to schedule a piece of work to a machine that already has the data and programs cached locally. The Grid MP caches data across all of the resources on the grid and uses data affinity scheduling to match jobs to resources that already have the data. This optimizes performance and minimizes network traffic across the grid.

Because resources on the grid are shared by all and are typically not dedicated to a single application, the grid provides features such as redundancy, monitoring, rescheduling, and checkpointing. Systems such as Seti@home and UD Grid MP allow the user to specify the number of copies of each piece of work that are to be scheduled. So long as one of the redundant copies completes execution, the job can complete. Another approach to overcoming a machine failure or unavailability is to detect that the machine is no longer available, and to reschedule the job to a different machine. This approach requires a way to monitor the state of the execution machine, detect a failure, and notify the scheduler to reschedule the job to a different machine. Systems such as Condor and UD Grid MP provide mechanisms to support application checkpointing that allows the job to be restarted on the same machine when it becomes available again, or the application can be rescheduled on a different machine and restarted using the checkpoint information.

Other than authentication of users, it is often a requirement that all data and programs be encrypted on the host machines. Systems such as the Grid MP provide automatic encryption/decryption of all input and output data, albeit at a

---

[5] LSF, *http://www.platform.com*

[6] PBS, *http://www.openpbs.org/main.html*

[7] Condor, *http://www.cs.wisc.edu/condor/*

[8] SRB, *http://www.npaci.edu/DICE/SRB/*

[9] Chimera, *http://www.griphyn.org/chimera/*

small performance cost. There are no source code modifications required to enable encryption and decryption of data.

13.5 CONCLUSION

Application integration can be achieved with little or no source code modification. Overall performance of an application is best when the resource requirements of the application match the resources available on the grid on which it is executed.

14 Grid-Enabling Network Services

David Daley

Montague River Networks, Inc.

In This Chapter

- Grid-Enabling Network Services
- OGSA Based Solution
- Optical LightPath Management

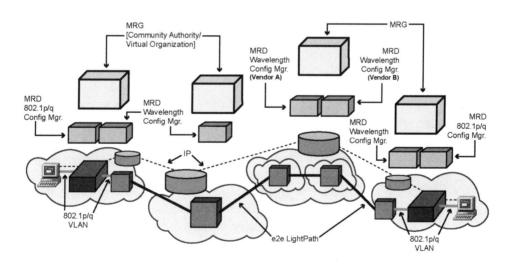

14.1 INTRODUCTION

The generic concept of differentiated services was the impetus for substantial telecom investment during the last decade. Certainly, when every service provider was offering identical network connection and transport services at cut rate costs the industry would falter, or so was the thinking at the time.

But surprisingly, after billions of dollars' investment and years of development, service offerings from the provider industry were still incredibly homogenous. In many ways this was to be expected, but to providers of network services it was a worrisome trend.

It was nearly impossible to charge customers premium rates for commodity services. As customers demanded lower and lower pricing, the artificial nature of the financial structure of the industry pushed providers to grab market share to the detriment of sustaining a reasonable business model. As could be predicted, the industry started to implode as fiscal reality once again set into the operating strategies of the providers.

This brings us back to differentiated services. Customers still won't pay premium rates for commodity networking. Providers still have the same infrastructural impediments that prevented services differentiation during the last decade.

Much of the problem has to do with the management systems for the networks themselves. The Operational Support Systems (OSS) that providers deploy are large, complex, and tightly coupled solutions that must be integrated internally amongst their components, and externally to both the devices in the network and the Business Support Systems (BSS) that drives the provider's operations.

The tight coupling of the layers of the solution creates the conditions that impede services differentiation. For a service provider to offer a new network-based service, they must first reengineer the OSS/BSS to support the service. This process can take anywhere from several months to a year or more; not exactly what one would consider responsive to rapidly changing market conditions.

The Open Grid Services Architecture (OGSA) provides a strategy for service providers to create service-oriented infrastructures which support more flexible resource management. Web Services supplies a paradigm that supports dynamic resource modeling, a fundamental requirement in pursuit of comprehensive management for evolutionary infrastructures. Peer-to-peer technology creates a mechanism for ad hoc relationships to be formed on demand, without a centralized controlling mechanism.

Montague River, leveraging the emerging grid, Web Services, and peer-to-peer standards, has developed technology which fundamentally shifts the way that network-attached resources can be managed. A service provider can now safely and securely supply to their customers fine grain control over arbitrary functionality contained within their network. Just as traditional grid technology has been deployed for years to allow users generic access to computational resources, Montague River's technology extends this capability to any network attached resource.

The service provider uses the technology to define the composition of their network and the types of low level and aggregate operations that their resources support. This creates a capabilities dictionary from which comprehensive services are composed. As the network evolves over time, the dictionary evolves to reflect the current reality.

Standardized user services are then composed from the supported operations and their definitions further extend the contents of the capabilities dictionary. Specialized services are used to describe the inventory and topology of the network. Thus, the solution not only supplies mechanism for defining network-based services, but also provides the functionality to then configure these services within the specific network itself.

Finally, the Montague River platform is OGSA-enabled. This means that any specific capability within the network can be exposed, complete with appropriate security mechanisms, to the grid and to grid-compatible applications.

Users of network resources now have a mechanism to define their own services, based on their own needs and requirements and these services can be deployed as required by the user. No longer is the choice between accepting what is available from third parties or building out one's entire network internally. Users can retain full control over their grid-based network services, even when they are brokered through external organizations.

Grid Service Providers (GSP) thus supply an enabling infrastructure which supports the user-driven services innovation of their customers, free from the delays associated with current infrastructural paradigms. When each user has the ability to innovate network services based on their own needs, the promise of the past decade will have arrived.

14.2 ON DEMAND OPTICAL CONNECTION SERVICES

Optical networking is one of the key technologies driving the growth of the networking industry. Optically based network devices have recently emerged as the connection device of choice for the user-facing edge of the service provider's network.

Users have realized that optical virtual private networks have the ability to bring together distributed organizations so that they can be efficiently restructured into cohesive logical groups as opposed to structures based on the geography of the office locations. However, organizations are faced with the traditional dilemma—

build out their own private optical network to gain control of the network and the ability to rapidly respond to changing market conditions or be faced with a service provider technology which requires centrally managed state for the creation of end-to-end optical connections.

Montague River's technology changes the nature of how the fulfillment of optical connection services occur. Providers now have more choice as to how they wish a state to be maintained, connections to be made, and services to be subsequently managed.

Providers can allow the state to be maintained at the edge of the network and can allow users the ability to autonomously and independently create end to end connections across their networks. Users can be empowered to cross connect and add-drop these connections independently of a central administrative organization. With this capability, users can then partition and re-advertise these connections to other users.

By plugging components of their networks into the grid, providers can join a global community of customer-controlled resources, allowing end users the ability to rapidly and efficiently pull together ad hoc communities of people and supporting infrastructure, glued together with on-demand optical connection services, and completed with the lack of a central administrator.

14.3 CREATING GRID-ENABLED NETWORK SERVICES

Montague River has developed a layered, component-based architecture, implemented as network management appliances, to deliver on the promise of grid-based resource virtualization. The solution implements the OGSA developed by the Globus Project, the peer-to-peer technology of the JXTA Project, and both SOAP and WSDL Web Services.[1, 2, 3, 4, 5]

Montague River leverages several open source technologies with our solution, including the Globus Toolkit 3, Apache Tomcat, AXIS, JBoss, MySQL, and Linux. By deploying such standardized implementations of key technologies, Internet scale compatibility for the solution is ensured.

---

[1] [Globus Project] Additional information available at *http://www.globus.org*

[2] Foster, Kesselman, Nick, Tuecke, *The Physiology of the Grid*, Infrastructure WG, June 2002.

[3] *Integration*, I. Foster, C. Kesselman, J. Nick, S. Tuecke. Globus Project, 2002. Available at *http://www.globus.org/research/papers/ogsa.pdf*

[4] [Project JXTA] Additional information available at *http://www.jxta.org*

[5] [Web Services] Additional information available at *http://www.w3.org/2002/ws*

The focus of the solution is toward ExtraGrids and InterGrids. These types of grids are evolutionary to the current Internet technology that connects people and organizations to each other. Distributed grids, such as those that connect enterprise partners together, require specialized support for policy management, security, accounting, and multi-site resource organization. When a grid extends to the global scale, regional, national and international issues must also be addressed.

Furthermore, users of global grids must be inherently shielded from the scale, complexity, and heterogeneity issues which are well beyond the control of their organization, if the grid is to become successful. This is the focus of standards organizations such as the Global Grid Forum and the World Wide Web Consortium. Similarly, this is the focus of Montague River's solution.

14.4 MONTAGUE RIVER GRID

Montague River Grid (MRG) supplies the necessary functionality to support the inter-domain and inter-provider management of network-based services.

MRG is a self-organizing grid adapter/gateway for network-attached resources and is deployed in conjunction with technology specific domain managers. MRG acts as the community authority/virtual organization for locally advertised and controlled network based services. Discovery, membership, registry, mapper, factory, notification, topology, and threading services are all supported.

Each MRG supports an aggregate capabilities dictionary from which domain-level capabilities are inherited and re-advertised. Furthermore, complex inter-domain services can be constructed and advertised as single service entities.

Once deployed, MRG enables user controlled, end-to-end, inter-domain and inter-provider services.

Components of the MRG include:

- Inter Domain Services—used to represent the persistent service datastore, service configuration, etc.
- Inter Domain Factory—primary entrance factory for users
- Factory—operational interface; used to implement the process of managing network services
- Registry—used to identify existing persistent service instances; inherited operational functionality of network devices
- Mapper—used to extract detailed information about existing service instances
- Notifications—used to relay asynchronous alarms and notifications from the network to the pertinent registered users of the affected resources

- Membership—used to enhance path selection within a business relationship or service paradigm
- Discovery—used to identify and propagate existing services within a business relationship or service paradigm

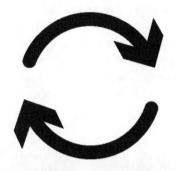

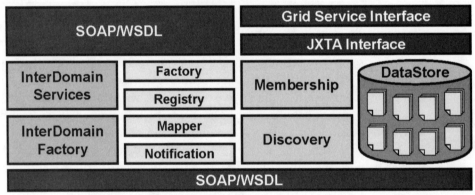

FIGURE 14.1 Montague River Grid architecture.

14.5 MONTAGUE RIVER DOMAIN

Montague River Domain (MRD) supplies the necessary functionality to support device-specific, domain-level management for network-based services.

MRD is a service fulfillment/configuration management platform for network-attached devices such as transport equipment, storage platforms, and computational servers. Its dynamically coupled network model allows network and service evolution without the re-engineering of the platform. Each MRD implements a capabilities dictionary from which component and comprehensive services are composed for the specific devices within its domain.

Furthermore, MRD implements standard functionality such as journaled transaction management, inventory upload and reconciliation, service configuration and rollback, service and topology reporting, and alarm correlation.

Components of the MRD as shown in Figure 14.2 include:

- Domain Services—service configuration operations, e.g., provisioning, service discovery, service grooming, etc.
- Domain Factory—network configuration operations, e.g., upload, transaction management, etc.
- Security—service and network security, including resource tagging, user enablement, etc.
- Configuration—specific service configuration operations, e.g, partitionLight-Path, findASPath, addXC, deleteXC, etc.
- Inventory—network device and component management, virtualized persistence of physical network, etc.
- Reporting—domain level network and service reporting.

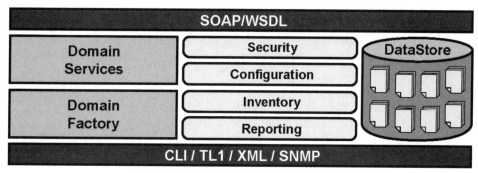

FIGURE 14.2 Montague River Domain architecture.

14.6 SAMPLE API

A full description of the API of the system is beyond the scope of this chapter. However, to give the reader a sense of the mechanisms of the solution, a couple of concepts are detailed below. The following descriptions use rough XML constructs as used by SOAP messages to describe the functionality.

The system makes heavy use of the `<criteria>` element. This is used to identify the objects under consideration during a specific action and which of the optional attributes a query will return via the API.

`<criteria>` is the standard query/matching construct used by the system. A match is made if the structure under inspection matches all of the regular expression (regexp) and comparison elements. This effectively performs a logical 'AND.' Most commands take an array of these objects to provide a logical 'OR.' There is also an optional `<not/>` element that, if provided, will invert the match. This is provided so that an existing query can easily be inverted without changing the logical expressions within the criteria.

Note that:

- `<regexp>` can be used against any data type, but `<dateTime>` values may not behave as expected.
- `<numComparison>` will only work against numeric values and will be ignored if non-numeric elements are specified.
- `<timeComparison>` will only work for date and time related values. It will be ignored if non-date elements are specified. The 2nd value will only be used by the 'between' comparison. The value used for 'before' will be the later and the value used for 'after' will be the earlier (that is, the expression is valid for both values).
- `<resultRestriction>` allows the builder of the criteria to limit the data in the element returned (if applicable). This mechanism results in less information being transferred over the wire and can have substantial performance improvements. If the element is absent, all of the elements of the return structure are included. If it is present, only the unique identifier of the object under consideration, the query elements, and the elements matching the provided XPath expressions are returned.

Although potentially complex, `<criteria>` elements will usually be fairly simple and will often identify only one or more id's or names.

```
<criteria> <!-- one or more of regexp and/or comparisons (simplifies
numeric matches)
                all must match, logical OR can be implemented via
multiple calls -->
        <not/> - optional
```

```
        <regexp>
            <element>
                <![CDATA[ XPath reference to the element in the target
XML  ]]>
            </element>
            <expression>
                <![CDATA[ A Perl 5 compatible regular expression ]]>
            </expression>
        </regexp>
        <numComparison>
            <element>...as for regexp...</element>
            <operation>gt | lt | eq | ne | gte | lte</operation>
            <value>...comparison value...</value>
        </numComparison>
        <timeComparison>
            <element>...as for regexp...</element>
            <operation>before | after | between</operation>
            <value>...comparison value...</value>
            <value>...2nd comparison value – used only by
'between'...</value>
        </timeComparison>
        <resultRestriction>
            <include>
                <![CDATA[ ... XPath expression indicating elements to
include in return
                data...]]>
            </include>
            ...
        </resultRestriction>
</criteria>
```

<report> is the definition of a report to be run. The <runReport> message actually executes the report to gather the data from the system. It defines a set of results to be drawn from service results and from node data. There may be zero or more of each of the <service> or <node> elements, in any order. The report results will be returned in a corresponding order.

A report is run against a set of input nodes. The entire node related criteria in the report template are selections against this subset rather than the complete set of nodes in the system (although an empty array being passed to the system will indicate to target all nodes).

Leaving out any node selection criteria will include the results for all nodes targeted by the report and thus will usually be absent.

The intent is to allow the referencing of an external data pump to post-process the data before returning them to the API. This is not to limit it to an XSLT translation, but rather to allow an external system to modify the data programmatically in any arbitrary fashion.

```
<report>
      <name>...</name>
      <description>...</description>
      <serviceData> <!-- 0 or more.  Retrieves data from service
execution; may occur in
                                any order interspersed with the other
elements-->
            <criteria>  <!-- 0 or more of these allowed -->
                <!-- criteria of a service for which to include results.
-->
            </criteria>
            <includeNode>  <!-- 0 or 1 of these allowed -->
                <criteria>  <!-- 1 or more of these allowed -->
                    <!-- criteria for nodes for which to include results
for this service.  (i.e.
                    include node results if a certain interface exist)
-->
                </criteria>
            </includeNode>
            <includeTemplate> <!-- include results of specific command
templates within
                                    service. Absence of this
element will include all -->
                <criteria>  <!-- 1 or more of these allowed -->
                    <!-- criteria of a command template for which to
include results.   -->
                </criteria>
                <data>  <!-- 0 or more of these allowed -->
                    <![CDATA[ ... XPath expression indicating actual
template element data to
                    include in return data...]]>
                </data>
            </includeTemplate>
      </serviceData>
```

```
        <nodeData>  <!-- 0 or more entries.  Retrieves data from the
model,
                            may occur in any order interspersed with the
other elements-->
        <criteria>  <!-- 0 or more of these allowed -->
            ... criteria executed against each targeted node.  Will be
used to determine if data
            for this node should be included. Include all if absent ...
        </criteria>
        <data>  <!-- 0 or more of these allowed -->
            <![CDATA[ ... XPath expression indicating actual node
element values to include in return data...]]>
        </data>
    </nodeData>
    <txData> <!-- 0 or more of these allowed -->
        <criteria>   <!-- 1 or more of these allowed -->
            ... criteria to selected the transaction...
        </criteria>
        <data> <!-- 0 or more of these allowed -->
            <![CDATA[ ... XPath expression indicating actual
transaction element data to
                include in return data...]]>
        </data>
    </txData>
</report>
```

<reportResult> is a formatted representation of the data being extracted from the system. This may be translated into a different format such as HTML, free text, or a different XML structure via an XSLT translation. the definition of the <report> element can be extended to include the XSLT or a reference to it.

Usually an XSLT will extract only the //reportResult/result/match/data/value for presentation, the remainder is included for cross-reference.

```
<reportResult>
    <result> <!-- There will be one of these created for each node for
each service, node,
                or 'tx' entry in the report template -->
        <type>serviceData | nodeData | txData</type>
        <match>
            <criteriaResult>
```

```
                    ... data from criteria used to match this entry — will
be a node, tx, or
                    service/template — restricted per resultRestrictions
in criteria ...
                </criteriaResult>
                <data> <!-- There will be one of these generated for each
data element in the
                        report -->
                <element>
                    ... <![CDATA[ ... Xpath expression indicating
which data element this
                        references ...]]>
                </element>
                <value>
                    ... the raw data extracted from the system ...
                </value>
            </data>
        </match>
    </result>
</reportResult>
```

14.7 DEPLOYMENT EXAMPLE: END-TO-END LIGHTPATH MANAGEMENT

An example of a forward-looking, user-controlled grid network service is the Light-Path. LightPath services are being pioneered by CANARIE Inc., and deployed in Canada's next generation Internet—the CA*net 4.

A LightPath service is defined to be: "Any uni-directional point to point connection with effective guaranteed bandwidth."[6] The first implementation of Light-Paths are across SONET add/drop multiplexers (Cisco Systems ONS 15454), but LightPaths can also be implemented over CWDM, DWDM, SONET, SDH, ATM CBR, MPLS LSR, DiffServ, and GigE VLAN devices and services.

Montague River has developed technology to support the user control of LightPath services with our MRG and MRD solutions.

---

[6] [LightPath Services] *User Controlled LightPaths Definition Document*, CANARIE Inc. Available at *http://www.canarie.ca/canet4/obgp/index.html*

Montague River's LightPath technology allows end users and grid applications to dynamically invoke spatial Quality of Service mechanisms to configure a dedicated optical BGP static route between two arbitrary points, independent of the number of service providers or types of network devices between the points. High-end data traffic is then automatically re-routed over the path. By setting up separate, direct optical BGP paths between source and destination, the use of advanced techniques to efficiently manage the transfer of large data sets can then be deployed.

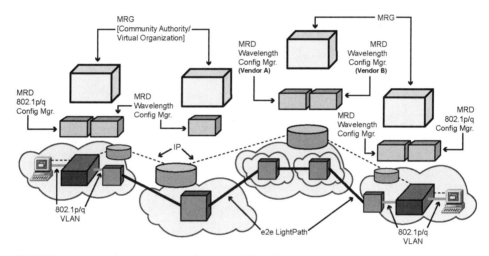

FIGURE 14.3   Deployment example—e2e LightPath management.

14.8 CONCLUSION

Grid Services provide a technology and a strategy for telecommunication service providers to create a service-oriented infrastructure which supports flexible resource management. This allows providers to offer new and differentiated service offering such as bandwidth on demand, customer-controlled wavelength services, and various types of virtual network services.

15 Managing Grid Environments

Daniel J. Feldman

GridFrastructure

In This Chapter

- Key Issues in Managing Grid Environments
- Defining Management Report
- Monitoring Grids
- Service Level Management

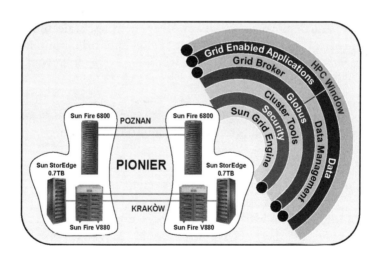

15.1 INTRODUCTION

Any implementation of new and complex technology necessarily entails new management challenges. Grid Computing is no exception. Substantial work on grid-enabling middleware technology has resulted in a variety of basic capabilities for secure and reliable resource sharing. In particular, the Globus Toolkit is a widely used and moderately mature collection of APIs and protocols that support the goals of intra-enterprise and inter-enterprise resource sharing. Globus is treated in depth elsewhere in this book, as is the utility model of Grid Computing that, by analogy with the electric power grid, leads to the name Grid Computing. The Globus Toolkit is widely deployed in the research community and is endorsed by a wide variety of commercial hardware and software vendors. The emergence of the Open Grid Services Architecture (OGSA) and release 3.0 of the Globus Toolkit should lead to an accelerating rate of adoption in commercial environments. This chapter will focus on the management requirements of commercial grid deployments.

15.2 MANAGING GRIDS

The traditional concerns of enterprise information technology managers apply as much to grid deployments as to any other enterprise scale information technology. To the extent that a grid deployment *crosses* enterprise boundaries, the manager's concerns increase substantially. We will examine the following broad categories of requirements:

- Trust—A grid deployment must achieve high levels of trustworthiness and the trust infrastructure must accommodate the unique needs for federated control.
- Management Reporting—The grid environment should provide a variety of reports that support management's need to understand the deployment and utilization of resources.
- Monitoring—A wide variety of processing tasks may be using the resources of the grid, and these need to be monitored in realtime and both normal and abnormal events must be reported using various modalities.
- Service Levels—Commercial collaborators who choose to use grid technologies have concrete expectations about service levels tied to specific business relationships; the tools and techniques for monitoring and managing service levels must be present.

■   Data Catalogs and Replicas—Sharing data resources is often the compelling motivation for the deployment of computing grids. Meta-data based mechanisms support the needs for data distribution in grids.

In addition to these management concerns, per se, facilities that support end user ease-of-use have an impact on the robustness of management tools. Although this chapter focuses on requirements rather than on any specific implementation, portal style user interfaces provide a useful integration point and so will be discussed briefly.

15.2.1 Trust

It is axiomatic that grid deployments must be trustworthy. The grid community has done substantial work ensuring that resources are accessible only to properly authorized and authenticated users. From the management perspective, it is important to understand this work in the context of a consistent trust model. A robust trust model has four major components: identity, privacy, authority, and policy.

15.2.2 Identity

The *identity* management problem has two aspects. First is the mechanism for asserting and assuring identity. Globus-based deployments, for example, rely on secure digital certificates and public key infrastructure to achieve reliable mutual authentication between users and resources. Second is the need to establish a regimen for assigning, accepting, and revoking identities. In commercial environments, individual identity is a delegation of the organization's identity. That is, the individual uses the firm's (and other firms') resources by virtue of his/her association with an employer. To the extent that s/he uses grid resources to accomplish a particular computing task, s/he is doing so in his/her capacity as an employee. As such, individual identities (and the trust associated with them) are extensions of the employing firm's identity. Consequently, the firm needs tools for issuing identities to users. In a PKI-based environment, these identities are represented by secure digital signatures. This means that the firm should either be operating as a certificate authority in order to issue and manage certificates or the firm must contract with an external certificate authority for those services.

All participants in a grid deployment must be able and willing to accept the certificates used to represent the identities of each other's employees. Normally, this is achieved by ensuring that either the certificate signer or the root certificate authority that certified the signer is identified as trusted in the appropriate control files. Additional confidence accrues if each externally certified identity is also required to solicit and receive explicit notification from the security authorities in outside organizations.

For example, assume that two companies, BigPharma, a global pharmaceutical company, and LittleBio, a start-up biotechnology firm, have decided to collaborate on a drug development project. The core task is to identify molecular targets for treating a specific disease and then to process BigPharma's library of possible drug molecules against the targets to see if one or more of them might work as a therapy. In this relationship, LittleBio is contributing special expertise in a particular biological system and both novel laboratory techniques and information processing techniques. BigPharma is contributing its library of compounds and substantial computing resources. Scientists from both firms will be routinely using each other's resources until the project is complete, and management has decided to deploy a Globus-based computing grid to facilitate this sharing. In this case, each firm wants to carefully control access to its highly valuable and proprietary resources. They mutually agree that only specific individuals that are contributing to the project are allowed access to those resources. In order to properly support this model, the identity management tools must work to accept or reject certificates based not just on the apparent validity of the signer, but also on specific individual identities.

In addition, it must be possible to suspend and revoke user's (represented by the previously approved external identity) access to the grid. Employees come and go from projects and leave the firm. The most useful identity management tools will provide at least a modicum of workflow support for the security administrator, for example, by presenting the administrator with a single, convenient point of access for receiving and acting on requests for changes in the state of individual identities.

15.2.3 Privacy

Establishing identity allows for reliable authentication. Authentication is the first step in a robust trust model. It allows the members and resources in the grid to trust that they are interacting with the member/resource that they intend. The next step is to ensure that the interaction itself is not compromised. That is, that no third party can use or misuse the substance of the interaction. Implementing encryption yields the desired privacy. Today's encryption technologies are sufficiently low-cost and computationally robust that large volumes of data can be transferred in an encrypted form without unacceptably degrading the system(s) performing the core computing tasks on the grid.

Academic and non-military government research environments generally eschew comprehensive encryption regimens. Because they are publicly supported efforts, often undertaken at national user facilities, they presume that the dictates of academic ethics and of collegial respect foreclose the necessity for such precautions. Further, they expect to publish final results and they often expect that intermediate results and raw data will be made public to support their publications. Clearly, commercial users have different expectations.

The technology-driven commercial enterprise derives much of its inherent value as an enterprise precisely because it achieves novel and unique results. Whether the result is the design for a new part, the identification of a new market segment, the location of new energy reserves or the structure of a new financial instrument, the cost to the firm of unauthorized and/or untimely disclosure or appropriation of computational results can be devastating.

Consequently, the commercial user of grid technology must pay careful attention to choosing an appropriate encryption regimen. The list below identifies aspects of the initiation, processing, and completion of a computational task that may each deserve to require the use of encryption:

- The specification of the parameters associated with a job
- The process of identifying a job to the grid and submitting the job for processing
- The transfer of the executable code associated wit the job from one location in the grid to another
- The transfer of input data required by the job from one location to another
- The transfer of output data produced by the job from one location to another
- The transfer of log files and control information generated as a byproduct of the processing job (e.g., stdout and stderr on Unix style operating systems)
- The notification of the user about the results and the delivery to the user of the results themselves
- Information that resides in the grid middleware and identifies the job and its outcome (e.g., internal log files)

Although some, none, or all of these may be important to the users of any particular grid deployment, it is up to the operators of and collaborators in each deployment to establish precisely what protections are necessary. Once the need is determined, it becomes a requirement for the middleware and the management tools to support that need.

15.2.4 Authorization

In addition to reliably established identities to authenticate users and encryption to ensure privacy, it is necessary to represent and enforce access controls. Requirements for access controls include granularity, federation, and policies.

Granularity refers to the ability to specify and enforce access controls on arbitrarily small discrete units of computational work. For example, BigPharma may only be willing to make a subset of its entire small molecule database available to researchers at LittleBio. For some researchers, this subset may be as small as a single compound while for others it is all the compounds in a particular family. LittleBio, in turn, may wish to restrict access to the results of specific experiments to management personnel in BigPharma while permitting unrestricted access to summary information to all BigPharma researchers. Clearly, the scope of the access control applied can vary substantially depending on the needs of the owner of the resource. The access control system must be able to adequately represent the full breadth of choices of scope that resource owners might want to make. Also, the access control system must be able to efficiently adjudicate requests for access.

Grid deployments that cross organizational boundaries have special requirements for distributed but complementary access controls. In particular, the specification of the access controls *must* be distributed in order to permit each of the organizations participating in the grid to implement an access control regimen that meets their specific needs. The adjudication of access controls may be distributed. In the distributed adjudication model, the adjudication request includes a representation of the requester's rights. Also, a resource owner may choose to delegate the specification (and perhaps the adjudication) of a subset for his/her resources to another member of the grid. This model of shared responsibility for the access control regimen is a *federated* model of access control. We believe that the minimum amount of federation necessary for a manageable grid deployment is the distribution of access control specification.

We use the term *policy* to represent the mechanism for specifying the mapping of access to subsets of the user base. An access control policy implicitly or explicitly associates a desired action (for example, read access to a database or the invocation of a particular job) with a set of resources and a set of users. The most flexible systems permit the set of users to be specified by including and/or excluding other sets of users. For example, one should be able to specify "all managers except managers in the Boston facility" or "all licensed professionals except Dr. Jones." The resources involved may be individually specified (for example, the policy for host1.abc.com) or specified as a group (e.g., abcHostList). In the simplest access control systems (and the ones most familiar to most users), file system access is controlled around a small set of possible actions: read, write, execute. In grid environments, however, this is insufficient for most resources. Although these file system actions are certainly an aspect of Grid Computing, a much larger and open-ended set of actions is more useful. For example, grid-specific functions (such as submitting a computational job, invoking a secure file transfer program, accepting an external credential, or requesting a system utilization report should all require appropriate authorization. Finally, the authorization policy should provide controls based on characteristics of the requesting context. Such characteristics include time of day and other date-related constraints, network IP address of the requester and, perhaps, the client application or Web browser that the user is using to access the system. A comprehensive implementation supports the extension of the policy mechanism to support site specific or currently unknown requirements.

Policy: The effectiveness of all security regimens ultimately reduces to the integrity of the policies and procedures implemented for and by people. Any organization contemplating the adoption of grid technologies should devote the necessary attention to articulating security policies. Policies should include, at a minimum, rules about the disclosure of private keys, procedures for reporting and disabling all credentials associated with lost and/or stolen keys, and periodic review of personnel lists and accepted credentials. Access control lists should be reviewed regularly, clear guidelines for inclusion on specific lists should be articulated, and policies regarding the exposure of the contents of the lists themselves should be developed. The Grid Computing environment should provide secure audit controls and appropriate reporting to enable (properly authorized) management personnel to validate the integrity of the operating environment.

15.3 MANAGEMENT REPORTING

A robust grid operating environment will collect a wide variety of metrics about the allocation and utilization of grid resources. Relevant metrics include both detailed and summary information about various categories of users and statistics on usage and utilization rates of various resources. This section itemizes a minimal set of reports that should be available (to appropriately authorized users) and the statistics that should be collected in order to construct the reports.

15.3.1 Users

A user report should contain the following information for each grid user, sorted by user id: # of requests submitted, # of requests succeeded, # of requests failed, # of requests cancelled, total CPU time used, total wall clock time used, total bytes transferred. The report should be available for a variety of time frames: today, week-to-date, month-to-date, year-to-date and, for an arbitrary time frame (within reasonable limits) daily, weekly, monthly, quarterly, yearly.

In addition, the report should be available—sorted by user within organizational unit *(Users by Organizational Unit)* with totals at organizational unit breaks and at organization breaks. Similarly, the same set of reports should be available— sorted by user within organization, with totals at organization breaks.

15.3.2 Resources

Two general types of reports are important for proper resource management. First are usage reports that are useful for understanding which users and/or organizations are using a particular resource or group of resources. Second are utilization reports that show the degree to which a resource or group of resources are approaching their capacity.

A resource usage report should contain the following for each grid resource, sorted by resource id: # of requests processed, # of requests succeeded, # of requests failed, # of requests cancelled, total CPU time used, total wall clock time used, total bytes transferred. The report should be available for a variety of time frames: today, week-to-date, month-to-date, year-to-date and, for an arbitrary time frame (within reasonable limits) daily, weekly, monthly, quarterly, yearly. The report should be available sorted by resource within organization with totals at organization breaks and sorted by resource list (if the management infrastructure allows resources to be grouped together in lists or some similar structure) with totals at resource list breaks.

It should be possible to request a report that shows all resource usage as a function of user requests. This report will show the same line item information as above, but will provide individual records and totals for each user who submitted a request that used the resources, including the option to sort and total by the user's organization and organizational unit.

Utilization reports should identify the resource (for example, BigPharma's supercomputer) and show the utilization of individual resources within the overall resource. Using the example of BigPharma's supercomputer, it would be reasonable to expect a report to track the percentage utilization of main memory, CPU(s), disks, major internal data structures (e.g., caches) and network bandwidth. As with usage reports, this information should be available for a variety of time frames and time periods. Unlike usage reports, a continuous utilization figure is being tracked, rather than discrete, user-initiated processing events. Consequently, the data collection and reporting mechanism should offer flexibility in the choice of measurement interval and duration. For daily reports, for instance, it may be useful to see utilization reported for every minute of the day. For weekly reports, it may be more useful to see hourly data. Often, this information is most usefully represented as a graph, rather than as a tabular report.

15.3.3 Jobs

A job detail report will identify each job run during a specified time period and, for each job, the user who submitted the job, the user's organizational unit and organization, whether or not the job succeeded and the resources used by the job. A variety of summary reports should be available that have job as the primary sort, but that total usage by user, by organizational unit, and by organization. As with the resource and user reports, this report (or set of reports) should be available for varying time frames and, within reason, historically over varying periods.

15.3.4 Audit Support

A key requirement for well-managed information systems is that they be auditable. It should be possible to identify and track all important events in the computing environment. Examples of important events include the submission of jobs, the completion of jobs, failed access attempts, and resource allocation failures. All actions that change the state of the grid itself—including the maintenance of user lists, the granting and revoking of credentials, the maintenance of access control lists and policies, and the creation and destruction of identity proxies—should be logged, and the logs should be available in the form of structure reports.

Clearly, any grid deployment that is achieving even modest utilization will quickly generate a substantial quantity of audit trail and log data. The reporting mechanism must be able to sort and filter the data in order to function as a useful management tool. Filtering should be available based on the type of event; all events that belong to a specific class (for example, `accessViolation` may be a class while `accessViolationSubmittingJob` would be a specific event type); events that are associated with specific users, organizational units, or organization; events that occurred during a specific time period; and events that are associated with a specific resource. Sorting and totaling should be flexible and support multiple levels of keys.

In addition, audit trails and logs must be routinely archived and readily available for inspection. The creation of back-ups themselves constitutes a loggable event, although it is common to use external (to the grid operating software) back-up utilities to implement a comprehensive data integrity and recovery program. In these circumstances, it may be difficult to ensure that back-up processing is integrated into the grid deployment as a logged or loggable event.

15.4 MONITORING

Computing grids are, by their nature, complex deployments of various hardware and software components. Systems and components will fail from time to time, as will individual processing jobs submitted by end users. Many events will be of sufficient importance that human interaction is promptly required. Consequently, the grid environment should provide a mechanism for real-time monitoring and notification. The notification mechanism should support a variety of modalities. Some users will prefer email notification while others will wish to be paged. Some users will base their decision about notification mode on their own judgment of the severity of each event while others will prefer to be notified when any event occurs and to be notified using all possible means.

15.4.1 Types of Events

Broadly, it may be desirable to provide notification of some sort for any event, including normal processing, that occurs as part of the operation of the grid. Some normal events, such as job completion, may be of sufficient importance and urgency that the initiator of the processing task that led to the event wants to know as soon as the job completes. Ideally, then, any event that could result in a log entry should be allowed to result in a realtime notification. Many logging subsystems support the concept of multiple event levels.

Some events are of only minor importance and get little or no attention when they occur, while others are considered critical to the proper operation of the grid and yield copious log entries. As with log entries, then, the event-monitoring subsystem may support multiple levels or types of event. For example, events can be categorized as *minor*, *moderate*, *serious*, or *urgent*. Then, a notification policy can be invoked based on the category associated with a particular event. Alternatively, a notification policy may be associated with each possible event, independent of that event's similarity to any other event.

15.4.2 Notification Modes

A notification policy may require notification via the most intrusive means (say, ringing a pager or telephone) for serious or urgent events, e-mail notification for moderate events, and no notification for minor events. A notification policy should be able to identify multiple responders, each associated with one or more notification modes, perhaps as a function of event type.

It may be useful to have a dynamically updated display of events. As events occur, the display is updated with the event and the current notification action being taken. Then, as personnel respond to the event, the event display is updated again to reflect the action taken. These updates can themselves be loggable events, creating an audit trail that is useful for tracking and reporting on the responsiveness of various service and support organizations.

15.5 SERVICE LEVEL MANAGEMENT

Whenever organizations share resources—whether they are organizations within the same enterprise or organizations from two or more cooperating enterprises—there is the expectation that each party is contributing according to some mutually agreed prior arrangement. Such arrangements may specify that a certain amount of processing capacity will be available during a given period of time, or that a certain amount of disk space and disk I/O will be available during that period of time, or that certain applications will be executed and will yield a specified throughput. In the utility or service grid model of distributed computing, it is crucial that resource providers are able to allocate costs to resource users and that resource users are able to hold providers accountable to specific service level commitments.

Much of the monitoring and reporting identified earlier in this chapter provide a solid foundation for the deployment of service-level management tools. Clearly, job reporting as described above provides the basis for verifying that known work loads are processed according to throughput commitments made by the resource provider. Summing job resource usage allows for a useful validation that total resources allocated as part of a grid deployment are consistent with resource provider commitments. Integrating this type of realtime usage and throughput information with a robust event management mechanism leads to the possibility of two important adaptive behaviors on the part of the grid infrastructure.

First, resource providers can use this information to automatically reconfigure or redeploy resources to a specific processing task or set of tasks. If, for example, a resource manager is throttling the arrival of jobs of a particular type (such as sequence alignment or data mining) and a queue length threshold is passed, a service-level management module could readjust the number of servers available for those jobs. Similarly, under appropriate circumstances, new services can be made available and underutilized services reduced.

Second, resource consumers can take advantage of realtime quality of service information to make service-level adjusted scheduling decisions. This is particularly useful in utility grid deployments. Given a choice of resource provider at the time that a processing task is presented to the grid, the scheduling mechanism (perhaps by means of explicit user choice) can search for the resource provider that best meets the user's service level requirement.

15.6 DATA CATALOGS AND REPLICA MANAGEMENT

Often, the motivation for a grid deployment is to enhance collaboration by making one or more valuable and (often) large data sets available to a community of users. It is necessary, then, to be able to identify the data set(s) that are to be shared and to be able to specify and monitor the distribution of those data sets among the user community.

15.6.1 Data Catalog

The data catalog is meta-data that identifies the data sets being managed. Typical meta-data includes the name of the data set, its location, the date-time it was last modified, its size, its type, and the correct access method. The catalog should be flexible enough to track everything from a subset of a file in a native operating system file format to a single record in a database manager to an entire database. Key issues in meta-data management include the completeness of the meta-data and the timeliness of updates to it. Catalog management should be subject to strict access controls and most catalog maintenance activity should be assiduously logged.

15.6.2 Replication

Although the data catalog describes the shared data resources, a *replication* mechanism manages the physical distribution of the data. Data can be either *pushed* or *pulled* to its destinations. The push approach is the traditional replication implementation. In this scheme, a replication manager, operating on a schedule of some sort, periodically initiates the movement of the data to their designated replication destination(s). The person administering the replication function must be able to specify the circumstances under which to initiate the replication. This should include at least a schedule-based specification mechanism; for example, "move dataset A (as understood by the catalog) to locations Alpha and Beta every weekday at 2am." Additional flexibility can be achieved by allowing the user to specify size-based criteria (e.g., when the file exceeds a particular size) or event based criteria (e.g., when job WeeklyUpdate completes successfully).

The pull approach responds to actual demand for access to the data. As data are requested across the grid, the replication mechanism makes copies available in at locations proximal to the requester. This *cache forward* approach has to deal with all the traditional sources of cache coherence, such as concurrent updates.

Whichever approach is used, the replication mechanism has to handle failed data transfer operations, unreachable network destinations, and unavailable source data (for example, in the event of a database manager failure).

15.7 PORTALS

As mentioned early in this chapter, Web-based applications that encapsulate grid operations and present a uniform user interface to the grid user base can be a useful integration point for all this management capability. For the average grid user, the portal provides a uniform user interface that masks the difference between the hardware and software resources available on the grid. For the administrative user, the portal provides a coherent application environment that uniformly enforces access controls and ensures consistent logging.

15.8 CONCLUSION

Grid deployments are potentially complex and difficult to manage. Appropriate management tools can address the needs of IT managers for control, transparency, accountability, and security. We have identified a substantial body of requirements that a grid management tool (or tools) must meet in order to provide true value to managers contemplating a grid deployment.

16   Grid Computing Adoption in Research and Industry

Wolfgang Gentzsch

Sun Microsystems, Inc.

In This Chapter

- Grid Computing Deployments in Academic and Research Institutions
- Grid Computing Deployments in Industry

Grid Computing Evolution

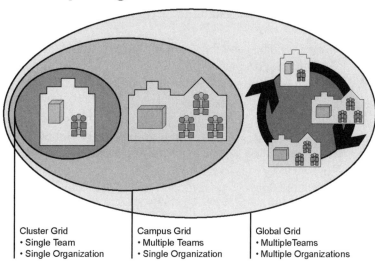

Cluster Grid
- Single Team
- Single Organization

Campus Grid
- Multiple Teams
- Single Organization

Global Grid
- MultipleTeams
- Multiple Organizations

16.1 INTRODUCTION

The Internet and the World Wide Web have improved dramatically over the past few years, mainly because of increasing network bandwidth, powerful computers, software, and user acceptance. These elements are currently converging and enabling a new global infrastructure called The Grid, originally derived from the electrical power grid which provides electricity to every wall socket. A computational or data grid is a hardware and software infrastructure that provides dependable, consistent, pervasive, and inexpensive access to computational capabilities, as described in Foster and Kesselman.[1] It connects distributed computers, storage devices, mobile devices, instruments, sensors, databases, and software applications.

This chapter describes some of the early grids we have designed and built together with our customers. In these projects, we have learned that grids provide many more benefits than just the increase in resource utilization, or using idle resources, as described in many articles today. Some of the key advantages a grid can provide are:

- Access—Seamless, transparent, remote, secure, wireless access to computing, data, experiments, instruments, sensors, etc.
- Virtualization—Access to compute and data services, not the servers themselves, without caring about the infrastructure.
- On Demand—Get resources you need, when you need them, at the quality you need.
- Sharing—Enable collaboration of (virtual) teams, over the Internet, to jointly work on one complex task.
- Failover—In case of system failure, migrate and restart applications automatically on another system.
- Heterogeneity—In large and complex grids, resources are heterogeneous (platforms, operating systems, devices, software, etc.). Users can choose the system that is best suited for their specific application.
- Utilization—Grids are known to increase average utilization from some 20 percent to 80 percent and more. For example, our internal Sun Enterprise Grid (with currently more than 7,000 processors in three different locations) to design Sun's next-generation processors is utilized at more than 95 percent, on average.

These benefits translate into high-level value propositions which are especially beneficial to upper management in research and industry who has to make the decision to adopt and implement a grid architecture within the enterprise. Such values are:

---

[1] Ian Foster, Carl Kesselman, *The GRID: Blueprint for a New Computing Infrastructure*, Morgan Kauffman Publishers, 1999.

- Increased agility—shorten time to market, improve quality and innovation, reduce cost, increase return on investment, reduce total cost of ownership
- Reduced risk—better business decisions, faster than competition
- Enabled innovation—develop new capabilities, do things previously not possible

Some of these benefits are present already in small managed-compute cluster environments, often called *departmental* grids. In fact, today, over 7,000 cluster grids are in production, running the Distributed Resource Management (DRM) software Sun Grid Engine,[2] or its open source version Grid Engine.[3] A few hundred of those early adopters are already implementing the next level, so-called campus or enterprise grids, connecting resources distributed over the university campus or the global enterprise, using the Sun Grid Engine Enterprise Edition.[2] And a few dozen of them are currently transitioning toward global grids, connecting resources distributed beyond university or enterprise firewalls, and using global grid technology such as Globus[4] and Avaki,[5] integrated with Sun Grid Engine. This strategy of evolutionary transition from cluster, to enterprise, to global grids is summarized in Figure 16.1

Grid Computing Evolution

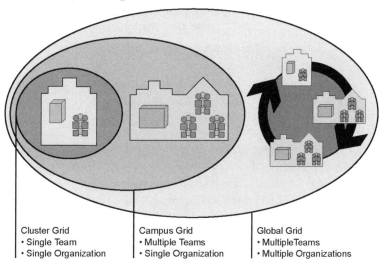

Cluster Grid
- Single Team
- Single Organization

Campus Grid
- Multiple Teams
- Single Organization

Global Grid
- MultipleTeams
- Multiple Organizations

FIGURE 16.1 Evolution of Grid Computing: cluster, enterprise, and global grids.

---

[2] Sun Grid Engine Web site, *http://www.sun.com/grid*

[3] Grid Engine open source project at *http://gridengine.sunsource.net/*

[4] Globus Web site, *http://www.globus.org*

[5] Avaki Web site, *http://www.avaki.com*

16.2 A GLOBAL GRID ARCHITECTURE

The Globus Project[6] introduces the term Virtual Organization (VO) as a set of users in multiple network domains who wish to share [some of] their resources. The virtual organization may be (as is currently the norm) a group of academic institutions that wish to enable resource sharing. The possible functional objectives of such resource sharing include:

- Increasing the throughput of users' jobs by maximizing resource utilization.
- Increasing the range of complementary hardware available, e.g., compute clusters, large shared memory servers, parallel computers, etc.
- Providing a (virtual) supercomputer grid which presents a platform for Grand Challenge applications.

The main software components (software tools) of a global compute grid are:

- User interface. Enabling remote, transparent, secure access to the grid resources by non-expert users, through some straightforward interface, usually a Web-portal.
- Broker. Automating job scheduling based upon the users' policies. Such policies could describe the users' priorities in terms of job requirements, available budget, time requirements, applications, etc. The broker would use these policies when negotiating on the users' behalf for a resource on the grid.
- Security, data-management, job-management, and resource discovery. These are the key issues that have been addressed by the Globus project.
- Resource guarantees and accounting. This is an area of current research activity and links in with the brokering technologies.

In the following section, we describe the components of a simple compute grid where geographically dispersed compute and storage resources are brought together and presented to users as a unified resource. Firstly, the general concepts are discussed and then, in the following sections, several specific grid implementations are described.

16.3 CORE COMPONENTS FOR BUILDING A GRID

To provide the grid functionalities and benefits described in the previous section, a set of core middleware components is necessary. In our grid projects, we are using, among others, the following components:

---

[6] Globus Web site, *http://www.globus.org*

- Access Portal—Grid Engine Portal, a few thousand lines of Java code for the Graphical User Interface, available in open source,[7] with the functionality to plug into any Portal Server, e.g., Sun ONE or Apache Portal Server, for additional security, authentication, authorization, and more.
- Globus Toolkit—The Globus Security Infrastructure (GSI), the Globus Resource Allocation Manager (GRAM), the Monitoring and Discovery Services (MDS), and GridFTP for enabling efficient file transfer.
- Distributed Resource Management—e.g., Sun Grid Engine, Sun Grid Engine Enterprise Edition, Condor,[6] LSF,[7] or PBS.[8]

16.3.1 Distributed Resource Managers

The core of any cluster or enterprise grid is the Distributed Resource Manager (DRM). Examples of DRMs are Sun Grid Engine,[2] Platform Computing's Load Sharing Facility,[7] or Altair's PBS Pro.[8] In a global compute grid, it is often beneficial to take advantage of the features provided by the local DRMs. Such features may include the ability to strongly complement the limited authorization currently available through Globus.

One of the key advantages of local DRM software is that it can simplify the implementation of the Globus layer above it. Specifically, where the underlying compute resources are heterogeneous in terms of operating platform, processor architecture, and memory, the local DRM provides a virtualization of these resources, usually by means of the queue concept.

Different DRMs have different definitions of a queue; but essentially a queue—and its associated attributes—represents the underlying compute resource to which jobs are submitted. If a Virtual Organization (VO) chooses to implement a specific DRM at each of its cluster grids, then the concept of implementing a virtualization of all the cluster grids is relatively straightforward, despite the possibility that the underlying hardware may be quite heterogeneous. One simply aggregates all the queue information across the VO. Because the attributes of the queues will have a common definition across the VO, the interface to this grid could be designed to be analogous to that implemented at the campus level (see the White Rose Grid in Section 16.4.5).

As an example of a DRM, Sun Grid Engine is a distributed resource management software, which recognizes resource requests and maps compute jobs to the

---

[2] Sun Grid Engine Web site, *http://www.sun.com/grid*

[3] Grid Engine open source project at *http://gridengine.sunsource.net/*

[6] Condor Web site, *http://www.cs.wisc.edu/condor/*

[7] LSF Web site, *http://www.platform.com/products/wm/LSF/index.asp*

[8] PBS Web site, *http://www.altair.com/pbspro.htm*

least-loaded and best suited system in the network. Queuing, scheduling, and prioritizing modules help to provide easy access, increase utilization, and virtualize the underlying resources. Sun Grid Engine Enterprise Edition, in addition, provides a Policy Management module for equitable, enforceable sharing of resources among groups and projects, aligns resources with corporate business goals via policies, and supports resource planning and accounting.

There are mainly two methods for integrating the local DRM with Globus:

- There is an integration of the DRM with GRAM. This means that jobs submitted to Globus (using the Globus Resource Specification Language, RSL) can be passed on to the DRM. Evidently the key here is to provide a means of translation between RSL and the language understood by the DRM. These are implemented in Globus using GRAM Job manager scripts.
- There is an integration with MDS. The use of a GRAM Reporter allows information about a DRM to be gathered and published in the MDS. The reporter will run at each campus site periodically via cron, and query the local DRM. This means that up-to-date queue information can be gathered across many departmental grids.

16.3.2 Portal Software and Authentication

The portal solution may be split into two parts: The first part is the Web server and/or container which serves the pages. Examples include Sun ONE Portal Server, Tomcat/Apache, uPortal. The second part is the collection of Java servlettes, Web services components, Java beans, etc., that make up the interface between the user and the Globus Toolkit and that run within the Server. The Grid Portal Development Kit,[27] is an example of a portal implementation that is interfaced with the Globus Toolkit.

16.3.3 The Globus Toolkit 2.0

In our early grid installations, we used Globus Toolkit 2.0 and 2.2 (GT2.0, GT2.2). The Globus Toolkit is an open architecture, open source software toolkit developed by the Globus Project. A brief explanation of GT2.0 is given here for completeness. Full description of the Globus Toolkit can be found at the Globus Web site.[4] GT3.0 re-implements much of the functionality of GT2.x but is based upon the Open Grid Services Architecture, OGSA.[9] In the following, we briefly describe the three core components of GT2.0 (and GT2.2).

---

[4] Globus Web site, *http://www.globus.org*

[9] OGSA Web site, *http://www.globus.org/ogsa/*

[27] Entrust secure Web portal, *http://www.entrust.com/solutions/webportal/*

16.3.3.1 Globus Security Infrastructure (GSI)

The Globus Security Infrastructure provides the underlying security for the Globus components. GSI is based upon Public Key encryption. Each time any of the components are invoked to perform some transaction between Globus resources in that VO, GSI provides the mutual authentication between the hosts involved.

16.3.3.2 Globus Resource Allocation Manager (GRAM)

GRAM provides the ability to submit and control jobs. GRAM includes the RSL Resource Specification Language in which users can describe their job requirements. Once submitted, the job may be forked on some grid resource or may be passed on to a DRM such as Condor or Grid Engine.

16.3.3.3 Monitoring and Discovery Services (MDS)

MDS provides the ability to discover the available resources on the grid. MDS implements a hierarchical structure of LDAP databases. Each grid resource can be configured to report into a local database and this information is aggregated in the higher level databases. Grid users can query the high level databases to discover up-to-date information on grid resources.

16.4 EXAMPLES OF RESEARCH AND INDUSTRY GRID IMPLEMENTATIONS

We follow mainly two ways of building a Global Grid infrastructure: either one starts with a local testbed, with a few systems, usually in one department (computer science, IT) and grows in the environment as confidence in the grid technology grows. Or one starts designing and building the complete grid architecture from scratch.

In the remainder of this chapter, we'll discuss a few typical real-life examples using either of these two approaches to build Grid Computing environments:

- The GlobeXplorer Grid—From departmental to global Grid Computing
- Houston University Campus Grid—Environmental modeling and seismic imaging
- HPCVL—the HPC Virtual Laboratory, connecting four universities in Ontario
- Canada NRC-CBR BioGrid—The National Research Council's Canadian Bioinformatics Resource
- White Rose Grid—Regional grid resources for the universities of Leeds, York, and Sheffield in the UK
- PROGRESS—Polish Research on Grid Environment on Sun Servers, combining grid resources of the universities in Cracow and Poznan

16.4.1 GlobeXplorer: From Departmental to Global Grid Computing

Because most Internet traffic to these sites originates in North America, site traffic patterns generally exhibit a regular diurnal pattern attributable to the users of the services: human beings. A biweekly graph of one of GlobeXplorer's application activity is shown in Figure 16.2.

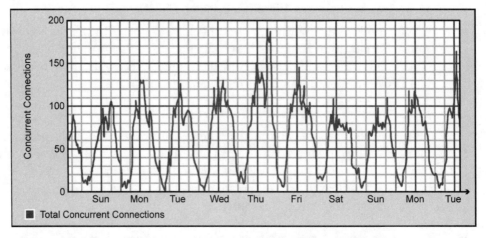

FIGURE 16.2   Site traffic pattern for GlobeXplorer application activity.

Organizations such as these often experience spikes in traffic attributed to fluctuations in user behavior from news events, promotional offers, and other transient activity. They must plan capacity to handle these spikes, which may last for perhaps an hour or two, and allow most capacity to become idle at night. Single-site Grid Computing typically comes as a welcome event to most Web-based companies, because they suddenly realize that at night they own a virtual supercomputer doing nothing, and various hypothetical projects thought impractical are now merely background jobs. Such was the case with GlobeXplorer, when it was faced with ingesting a national archive of several hundred thousand frames of raw aerial photography, each requiring nearly 30 minutes of dedicated CPU time[10] to make it usable on the Internet. Because of the loose-grained parallelism of the core jobs, Sun Grid Engine (SGE) provided an ideal mechanism to address this problem. Almost overnight, using SGE, the ingest process was transformed from being CPU-bound to I/O bound: GlobeXplorer couldn't feed "the beast" fast enough.

---

[10] This processing involved geometric corrections for terrain distortion, color enhancements, and wavelet compressions.

At GlobeXplorer,[11] basic precautions were taken from the beginning to ensure that SGE jobs prototyped in the lab were effectively isolated in production in terms of directory structures, prioritization, and permission. Supporting database schemas and workflow have stabilized to the point that overall utilization of the production machines has increased from 43 percent to over 85 percent, and avoided nearly $750,000 in capital expenditures and operating on this project alone. Processing several terabytes of imagery each week along side production applications is now a routine phenomenon.

A side effect of full utilization is that site resources are now mission-critical to more than one business process: in-line http traffic and content ingest. Intuitive observations of the compute farm statistics must now account for the fact that requests originate from both the Web and the SGE scheduler, and take into account several workflows overlaid on the same compute, networking, and storage resources. Configuration management issues concerning system library versioning, mount points, and licensing must also make these same considerations.

Because of the separation of the front office (the http compute farm) and the back office (workflow controller), multisite coordination of scheduling quickly became an issue. It was perceived that this was precisely what Globus was designed to address, which complemented the functionality of SGE ideally. Due to the tremendous success of the initial SGE-based implementation, executive management at GlobeXplorer has embraced the concept of Grid Computing and supports the increased use of grid technologies throughout the enterprise.

If one looks at a map, the data portrayed clearly originate from multiple sources (population data, street networks, satellite and aerial archives, etc). Globus is now perceived as a framework for supply-chain management of such complex content types that will ultimately become a part of GlobeXplorer's product offerings. Experimentation is already underway with remote data centers that have very specialized image algorithms, larger content archives, and computing capacity constructing such a supply chain, as shown in Figure 16.3.

Such chains might include coordinate reprojection, format transcodings, false-color look-ups for multispectral data, edge enhancements, wavelet compression, buffering zone creation from vector overlays, cell-based hydrologic simulations, etc. Routines for these operations will exist in binary forms at various locations, and will require various resources for execution. Clearly, it will often be far more efficient to migrate an algorithm to a remote archive than it will be to migrate an archive to an instance of an algorithm. GlobeXplorer plans to remotely add value to previously inaccessible/unusable archived data stores by using OGSA/WSDK v3 to provide appropriate product transformation capabilities on

---

[11] GlobeXplorer Web site, *http://www.globexplorer.com*

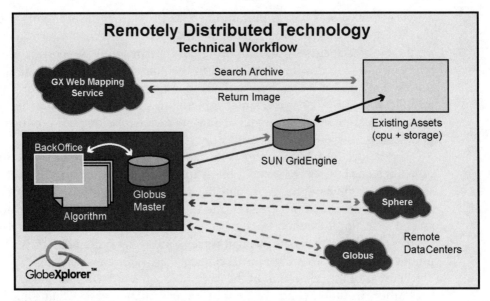

FIGURE 16.3 Globus as a framework for a supply-chain management grid.

the fly appropriate to a transient supply chain and to create derivative product offerings, with guaranteed service level agreements (SLAs), to the mission-critical applications of our customers.

16.4.1.1 Meta-data and Registries

This vision of chained flows of discoverable data and services is shared by other participants in the standards body of geographic information: the OpenGIS consortium,[12] as shown in Figure 16.4.

Chains themselves will be expressible in emerging standards such as WS-Route, WSFL, and related service-oriented workflow description languages, expressed as trading partner agreements (TPAs) within ebXML registries.

Perhaps the biggest challenges to making this vision a reality are standards-based registration and discovery issues and meta-data harmonization across disparate, often overlapping standards at the organizational, local, regional, and national levels, implemented in often overlapping technologies such as LDAP and Z39.50.[13] These issues are actively being addressed in the context of pressing applications such as homeland security, which requires immediate access to critical,

---

[12] OpenGIS Web site, *http://www.opengis.org*

[13] OpenLDAP Web site, *http://www.openldap.org*

often sensitive data, regulated by federal freedom of information and privacy legislation, network security, and privacy issues (public vs. private data).

Internet companies—such as GlobeXplorer—that have already adopted a cluster-based approach to large http traffic load are ideal candidates to adopt grid technologies, because they already have appropriate hardware, networking, and operational support infrastructure. When opportunity and/or necessity knocks at their door, competitive organizations will view their options with grid technologies in mind.

16.4.2 Houston University Campus Grid for Environmental Modeling and Seismic Imaging

Several major research activities at the University of Houston (UH) require access to considerable computational power and, in some cases, large amounts of storage. To accommodate many of these needs locally, UH decided to create and professionally operate a campus-wide grid that combines this facility with departmental clusters using Grid Engine for job submission. Future plans include collaborating with regional and national partners to form a wide area grid.

A variety of ongoing scientific research projects at UH require access to significant computational resources. Researchers in chemistry, geophysics, me-

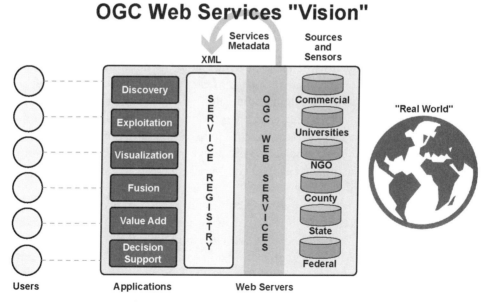

FIGURE 16.4 OGC Web Services vision.

chanical engineering, computer science, and mathematics are among those who routinely make use of parallel and clustered systems across campus. One of the most demanding collaborative research efforts involves a team of scientists who are working on the numerical simulation and modeling of atmospheric pollution, with a special focus on subtropical Gulf Coast regions such as Houston-Galveston-Brazoria. Their work includes developing and deploying a parallel version of a community air quality model code that will forecast the impact on atmospheric pollution of various strategies for the reduction of volatile organic compounds and nitric oxide. A photochemical air quality model consists of a set of coupled partial differential equations, one for each chemical species. The input to these equations is the complete local weather data and concentrations of chemical precursor molecules, ideally from realtime monitoring. The execution scenario for this project therefore involves the execution of a limited area weather model—indeed, multiple limited area models are executed on increasingly smaller, but more closely meshed, domains—in conjunction with the chemical model. It also relies on global weather data that are automatically retrieved each day. The grid system is required to start the weather model once these data are locally available and, once the weather code has reached a certain phase, initiate execution of the chemical code. As these are run at separate grid sites, the files must be automatically transferred.

Geophysicists at UH are engaged in research, development, and evaluation of seismic processing and imaging algorithms, which generally involves the handling of very large data sets. Geophysicists use the systems described below to develop methods for *high resolution imaging of seismic data* in order to better identify and quantify hydrocarbon reserves and aggressively disseminate results to the oil and gas industry. Industry partners provide real-world data and the opportunity to verify results via testing in a working oil field. Work includes developing a 3-D pre-stack wave equation depth migration algorithm. (3-D pre-stack imaging is the most compute intensive application currently run within the super major oil and geophysical service companies, consuming the vast majority of CPU cycles.) The workload generated by these scientists includes both low-priority long-running jobs and short, high-priority imaging algorithms. High-speed access to the storage system is critical for their execution.

UH has both large and small computational clusters connected through optical fiber across the campus. The hardware available at the High Performance Computing Center (HPCC) includes a cluster of Sun Fire 6800 and 880 platforms connected via Myrinet. They are available to a broad cross-section of faculty and are deployed for both research and teaching. This facility is heavily utilized, with a high average queue time for submitted jobs. We therefore exploit the availability of other resources on campus to alleviate this problem by operating a number of different systems including those at the HPCC in a campus-wide grid.

The campus grid, Figure 16.5, is divided into several administrative domains corresponding to the owner of the hardware, each of which may contain multiple clusters with a shared file system. In our environment all basic grid services—such as security and authentication, resource management, static resource information, and data management—are provided by the Globus Toolkit,[1, 14] whose features may be directly employed by accredited users to submit jobs to the various clusters. An independent certification authority is managed by HPCC. The Sun Grid Engine (SGE[2, 3]) serves as the local resource manager within domains and is thus the software that interfaces with the Globus resource manager component. However, many application scientists find dealing with a grid infrastructure a daunting task. UH has therefore developed a portal interface to make it easy for them to interact with grid services. It can be used to obtain current information on resources, move files between file systems and to track their individual account usage and permissions, as well as to start jobs.

The locally developed EZ-Grid system,[15] using Globus as middleware, provides an interface to authenticate users, provide information on the system and its status, and to schedule and submit jobs to resources within the individual domains via Grid Engine. The development of EZ-Grid was facilitated by the Globus CoG Kits,[16] which provide libraries that enable application developers to include Globus's middleware tools in high-level applications, in languages such as Java and Perl. The portal server has been implemented with Java servlets and can be run on any Web server that supports Java servlets.

Users can access grid services and resources from a Web browser via the EZ-Grid portal. The grid's credential server provides the repository for the user credentials (X509 certificate and key pairs and the proxies). It is a secure standalone machine that holds the encrypted user keys. It could be replaced by a MyProxy server[17] to act as an online repository for user proxies. However, this adds an upper

---

[1] Ian Foster, Carl Kesselman, *The GRID: Blueprint for a New Computing Infrastructure*, Morgan Kauffman Publishers, 1999.

[2] Sun Grid Engine Web site, *http://www.sun.com/grid*

[3] Grid Engine open source project at *http://gridengine.sunsource.net/*

[14] I. Foster, C. Kesselman, S. Tuecke, "The Anatomy of the Grid: Enabling Scalable Virtual Organizations," *International Journal of Supercomputer Applications*, 15(3), 2001.

[15] B. M. Chapman, B. Sundaram, K. Thyagaraja, "EZ-Grid System: A Resource Broker for Grids," *http://www.cs.uh.edu/~ezgrid*

[16] G. von Laszewski, I. Foster, J. Gawor, W. Smith, and S. Tuecke, "CoG Kits: A Bridge between Commodity Distributed Computing and High-Performance Grids," ACM 2000 Java Grande Conference, 2000.

[17] J. Novotny, S. Tuecke, V. Welch, "An Online Credential Repository for the Grid: MyProxy." *Proceedings of the Tenth International Symposium on High Performance Distributed Computing (HPDC-10)*, IEEE Press, August 2001.

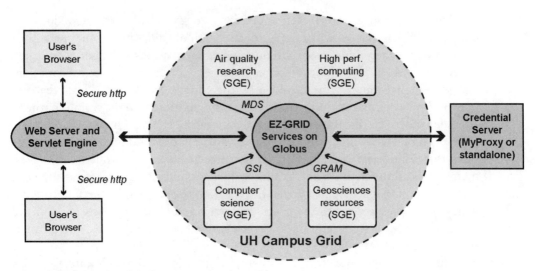

FIGURE 16.5 University of Houston campus grid.

limit to the mobility of the users due to the limited lifetime of the Globus proxies delegated. UH has adopted the standalone credential server model to allow users to access grid services with unlimited mobility through a browser even if they have no access to their grid identities. Appropriate mapping of the portal accounts to proxies allow users to perform single sign-on and access grid resources. The portal supports the export of encrypted keys from the user to the credential server, using secure http sessions. In some scenarios, the credential server can also be used to generate the user credentials, thus ensuring enhanced security.

EZ-Grid provides the following major grid services for its users:

- *Single sign-on*—Globus proxy (temporary identity credential) creation using Grid Security Infrastructure[18] and X509 certificates. This allows the user to seamlessly establish his or her identity across all campus grid resources.
- *Resource information*—Viewable status information on grid resources, both static and dynamic attributes such as operating systems, CPU loads, and queue information. Static information is obtained primarily from Globus Information services such as MDS^2 and dynamic scheduler information and queue details are retrieved from SGE. Users can thus check the status of their jobs, load on the resources, and queue availability. Additional information provided in-

[2] Sun Grid Engine Web site, *http://www.sun.com/grid*

[18] I. Foster, C. Kesselman, G. Tsudik, S. Tuecke, "A Security Architecture for Computational Grids," *ACM Conference on Computers and Security*, 1998, 83–91.

cludes application profiles (meta-data about applications), job execution histories, and so forth.

■ *Job specification and submission*—a GUI that enables the user to enter job specifications such as the compute resource, I/O, and queue requirements. Automated translation of these requirements into Resource specification language (RSL)[19] and subsequent job submission to Globus Resource Allocation Managers (GRAM)[20] are supported by the portal. Scripts have been implemented to enable job handoff to SGE via Globus services. Further, automated translation of some job requirements into SGE parameters is supported.

■ *Precise usage control*—Policy-based authorization and accounting services[21] to examine and evaluate usage policies of the resource providers. Such a model is critical when sharing resources in a heterogeneous environment such as the campus grid.

■ *Job management*—Storage and retrieval of relevant application profile information, history of job executions, and related information. Application profiles are meta-data that can be composed to characterize the applications.

■ *Data handling*—Users can transparently authenticate with and browse remote file systems of the grid resources. Data can be securely transferred between grid resources using the GSI-enabled data transport services.

Figure 16.6 shows how EZ-Grid interacts with other middleware tools and resource management systems.

16.4.2.1 Conclusions

Public domain software such as Globus and Sun Grid Engine may be used to construct a grid environment such as the campus grid we have described. However, simplified access to grid services is essential for many computational scientists and can have a positive impact on the overall acceptance of this approach to resource utilization. A variety of projects—including HotPage,[22] Gateway, and UNI-CORE[23]—provide a portal interface that enables the user to access information

---

[19] Resource Specification Language, RSL, *http://www.globus.org/gram/rsl_spec1.html*

[20] K. Czajkowski, I. Foster, N. Karonis, C. Kesselman, S. Martin, W. Smith, S. Tuecke, "A Resource Management Architecture for Metacomputing Systems," Proc. IPPS/SPDP '98 Workshop on Job Scheduling Strategies for Parallel Processing, 1998.

[21] B. Sundaram, B. M. Chapman, "Policy Engine: A Framework for Authorization, Accounting Policy Specification and Evaluation in Grids," 2nd International Conference on Grid Computing, Nov 2001.

[22] J. Boisseau, S. Mock, M. Thomas, "Development of Web Toolkits for Computational Science Portals: The NPACI HotPage," 9th IEEE Symposium on High Performance Distributed Computing, 2000.

[23] Uniform Interface to Computing resource, UNICORE, *http://www.unicore.de*

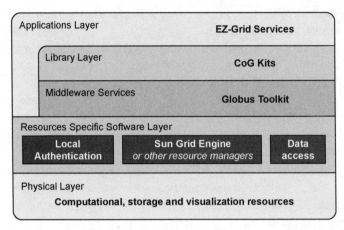

FIGURE 16.6 EZ-Grid services interaction with tools and resource management.

provided by Globus or the underlying grid services. Toolkits such as GridPort[24] and GPDK[25] have been developed to simplify the construction of portals that exploit features of Globus. Although the UH system is quite similar to these latter efforts, it provides more extensive functionality, in particular by exploiting information that can be easily retrieved from Sun Grid Engine also.

16.4.3 Ontario HPC Virtual Laboratory

The High Performance Computing Virtual Laboratory (HPCVL)[26] was formed by four Ontario universities (Carleton University, Queen's University, Royal Military College, and the University of Ottawa) to enable computational research in a variety of emerging research areas important to Ontario's and Canada's economy. Researchers who are pioneering globally significant discoveries will, through HPCVL (Ontario and Canada's premiere multidisciplinary facility for computational research), conduct pioneering and innovative research in areas such as population health, photonics, psychology, economics, nuclear physics, civil engineering, nanomaterials, applied mathematics, bioinformatics, and massive data handling. To date, more than 80 research groups involving over 300 researchers have been using HPCVL resources to conduct innovative research.

Through the recruitment of faculty and graduate students, HPCVL is helping each of the member institutions meet its strategic research objectives by providing state-of-the-art computing resources to researchers. HPCVL is also facilitating es-

---

[24] GridPort, *https://gridport.npaci.edu/*

[25] The Grid Portal Development Kit Web site is at *http://doesciencegrid.org/projects/GPDK/*

[26] HPCVL Web site is at *http://www.hpcvl.org*

tablished research at the partner institutions and research at institutions across Ontario and Canada. An example is the work of the Sudbury Neutrino Observatory (SNO) whose work was recently selected by *Science* magazine as the second most important discovery of 2002. HPCVL currently provides basic HPC calculation, data storage, and archive capability for SNO. HPCVL allows innovative researchers from across Canada to gain access to professionally run, state-of-the-art, secure HPC resources. HPCVL provides training in HPC techniques, and supports graduate students through a fellowship program.

HPCVL is developing and deploying a leading-edge 1024-bit encrypted secure Web-based portal to provide an easy-to-use secure interface that will protect intellectual property and other confidential information. This innovative solution will provide researchers from any location secure access to the resources they need to conduct their ground-breaking research and development.

HPCVL has embarked on an ambitious effort to make its resources available to researchers with sufficient security for the protection of intellectual property and confidential information, such as patient data, without introducing complexity. At the first level, HPCVL has implemented a Virtual Private Network (VPN) between the four founding member institutions, thus protecting information transmitted between the four sites as shown in Figure 16.7. The HPCVL production environment now consists of the main Sun cluster located at Queen's University, and Sun 6800s at both Carleton University and the University of Ottawa. A master Sun Grid Engine Enterprise Edition, combined with Sun Cluster Tools under Solaris 9, is utilized to perform resource management so that jobs submitted are run at the site that will give the most efficient use of the available resources.

The HPCVL portal effort is presently under development and scheduled for deployment in late summer or early fall. This effort entails the interfacing of software from Sun ONE, Sun Grid Engine Portal, Entrust Secure Web Portal Solution,[27] and elements of the Globus Toolkit. Sun Grid Engine Enterprise Edition, Solaris 9, and Sun ClusterTools[28] will still be utilized to manage the resources of the production environment. This effort will mean that researchers will be able to securely access HPCVL resources without worry about the complexity or the details of the solutions implemented to protect their research. The portal will help bring secure mobility to the researchers and enhance the tailoring of services to their needs.

HPCVL is seeking to upgrade the present installed infrastructure through applications to various government agencies and also seeks to form partnerships with private sector partners. These upgrades would eventually see the replacement of UltraSPARC III technology with UltraSPARC IV technology at the central site (Queen's University) and at the workup facilities in place at Carleton University

---

[27] Entrust secure Web portal, *http://www.entrust.com/solutions/webportal/*

[28] The source for Sun HPC ClusterTools can be downloaded from *www.sun.com/solutions/hpc/communitysource*

HPCVL

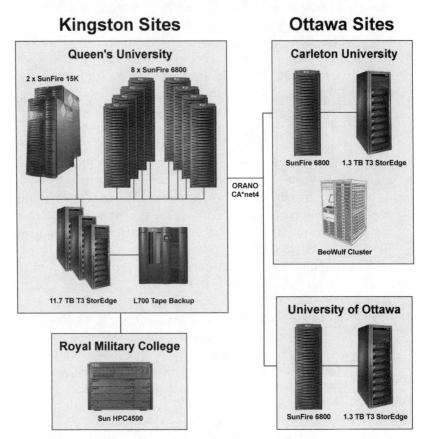

FIGURE 16.7 HPCVL grid connects four sites in Ontario.

and the University of Ottawa. In addition, the Beowulf cluster at Carleton will be upgraded. The central site provides researchers with world-leading computing environment in a production environment. The Beowulf cluster allows computer scientists and engineers to work on HPC systems, software, and tools in a non-production environment. The storage capacity at the central site will be expanded to accommodate the rapidly growing demand for both medium- and long-term storage of research data. With the desired upgrades complete, the central site at Queen's University will be the one of the top academic research HPC sites in North America, and one of the top 50 in the world.

16.4.4 Canada NRC-CBR BioGrid

The National Research Council of Canada's Canadian Bioinformatics Resource (NRC-CBR[29]) is a distributed network of collaborating institutes, universities, and individuals across Canada dedicated to the provision of bioinformatics services to Canadian researchers. NRC-CBR is also a Sun Center of Excellence[30] in Distributed BioInformatics. Some highlights of NRC-CBR include: Largest installation of the Sequence Retrieval System (SRS[31]) in North America; official European Molecular Biology Network (EMBnet)[32] Node for Canada; founding and active member of the Asian Pacific Bioinformatics Network (APBioNet[33]); North American node for ExPASy proteomics server.[34] NRC leverages the excellent high bandwidth of Canadian networks, namely CANARIE Inc.'s CA*net4,[35] to integrate data storage and standalone applications at member sites with centrally maintained databases and dedicated hardware to provide users with an environment that evolves with their needs and scales smoothly as membership grows.

NRC-CBR is conceived as a collaboration of peers, each contributing the unique expertise of their organization under a common umbrella. Thus, Web Services are distributed across member organizations, datasets, and applications developed at member sites are distributed through NRC-CBR, and hardware resources are donated by members for the common benefit, all underwritten by the funding of core services by the National Research Council of Canada.

Currently, NRC-CBR is grid-enabling servers across the country to provide a closer integration of member sites. In its bid to develop a Bioinformatics Grid for Canada, NRC-CBR is employing Cactus,[36] Sun Grid Engine (SGE), and Globus in collaboration with CANARIE, C3.ca,[37] and other NRC institutes. These collaborative efforts have formed the basis of a Memorandum of Understanding (MOU) between NRC, CANARIE, and C3.ca to form Grid Canada.[38] Grid Canada, through NRC, will collaboratively interconnect the HPC sites of more than 30 universities and NRC, including NRC-CBR's HPC.

---

[29] Canada NRC-CBR, *http://cbr-rbc.nrc-cnrc.gc.ca/*

[30] Sun Center of Excellence program, *http://www.sun.com/products-n-solutions/edu/programs*

[31] SRS Sequence Retrieval System, *http://srs.ebi.ac.uk/*

[32] European Molecular Biology Network, EMBnet, *http://www.embnet.org/*

[33] Asian Pacific Bioinformatics Network, APBioNet, *http://www.apbionet.org/*

[34] ExPASy Molecular Biology Server, *http://us.expasy.org/*

[35] CANARIE, *http://www.canarie.ca/about/about.html*

[36] Cactus problem solving environment, *http://www.cactuscode.org/*

[37] C3.ca, Canadian High Performance Computing Collaboratory, *http://www.c3.ca/*

[38] Grid Canada, *http://www.c3.ca/*

Cactus developers will collaborate with NRC-CBR to develop bioinformatics and biology thorns for integration with Cactus. Perl and JAVA will be integrated into the Cactus core, as bioinformatics applications are typically written with these languages. Integration of Cactus, SGE, and Globus components will be necessary to facilitate diverse functionality in applications using distributed multi-vender platforms.

Although new experimental approaches permit whole-organism investigations of proteomics, metabolomics, gene regulation, and expression, they also raise significant technical challenges in the handling and analysis of the increased data volume. NRC-CBR provides biologists across Canada with access to bioinformatics applications and databases, large-volume data storage, basic tutorials, and help desk support. CBR provides this service to scientists at the National Research Council of Canada as well as to academic and not-for-profit users associated with Canadian universities, hospitals, and government departments. Any infrastructure serving this community must provide a low-cost, secure, and intuitive environment integrating a wide range of applications and databases. From an administrative perspective, the solution must also be scaleable to accommodate increasing usage and data flow, and also promote economies of scale with regard to systems administration and user support. Over the last year or so, it has become clear that many of these requirements are served by emerging grid technologies. NRC-CBR grid architecture is shown in Figure 16.8.

In correlation with the grid developments, NRC-CBR is developing secure portal access to its services to effectively present users with a single sign-on secure environment where the underlying national grid infrastructure and distributed NRC-CBR resources are transparent to the user community. Biologists are only concerned with using the tools and accessing the data, not the complexities of the underlying grid or informatics architecture.

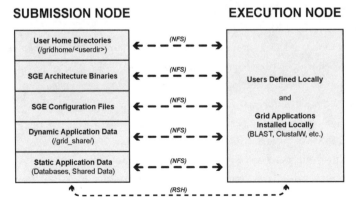

FIGURE 16.8 Initial NRC-CBR Global Grid architecture.

CBR is in the early stages of upgrading existing centralized infrastructure, forming a bioinformatics grid to integrate geographically distributed computational and data resources. An initial step toward this goal has been the formation of a CBR WAN cluster grid, linking member sites using the Sun Grid Engine Enterprise Edition software. As Figure 16.8 shows, the initial configuration did not establish a clear division between data, administrative, and execution elements of the grid architecture, and NFS mounting was used as the sole means of making data and binaries available on grid nodes, with network latency slowing overall performance. In the absence of NIS+[39] (due to conflicting local and grid user definitions) or a shared repository of application binaries, grid administration became a significant and increasing overhead. In order to ensure that all results were returned to the same location, the grid submission command *qsub* was wrapped in a script to set the output destination to the NFS-mounted user's home directory.

A range of bioinformatics applications are being made to interface with the CBR grid for eventual deployment. These applications fit into two categories, the first containing heavily used small utilities, which benefit from the scheduling features of a grid to spread load and parallelize workflow. As an example, most of the utilities in the European Molecular Biology Open Software Suite (EMBOSS[40]) fit well into this category. In the second category are applications, which can be parallelized but which are tolerant of the concurrency issues inherent in the loosely coupled distributed environment of a grid. Examples include any database search, such as the Sequence Retrieval System (SRS) (provided the databases to be searched can be subdivided and maintained on many nodes), advanced pattern matching searches such as BLAST® (Basic Local Alignment Search Tool) or Patternmatcher, or multiple sequence alignment where the algorithm is based on many pairwise alignments, such as ClustalW.[41] In conjunction with these developments, CBR plans to implement a revised grid architecture which more closely fits the requirements of low maintenance costs and a heterogeneous environment, as shown in Figure 16.9. The most obvious change is that the principal elements of grid function have been separated; administration and configuration are largely restricted to the Master node, which performs no other grid function. The rsync remote filesystem synchronization process will be used to ensure all grid nodes securely share a common configuration and appropriate OS and architecture-dependent application binaries. Although CBR makes extensive use of the CA*NET4 high-speed Canadian research network, stretching 5,500 miles from Halifax to Vancouver, not all member sites have high-bandwidth connections, and performance can be improved by

---

[39] Network Information Service Plus, *http://www.eng.auburn.edu/users/rayh/solaris/NIS+_FAQ.html*

[40] EMBOSS, The European Molecular Biology Open Software Suite, *http://www.hgmp.mrc.ac.uk/Software/EMBOSS/*

[41] ClustalW sequence analysis, *http://www.ebi.ac.uk/clustalw/*

adopting a mixed strategy of rsync for transfers of large or static datasets, and NFS (tunneling through Secure Shell for security[42]) where datasets are small or dynamic. Experience has also shown us that grid execution nodes frequently serve other non-grid functions at member sites, and therefore we will initially provide Network Information System (NIS+) user authentication to supplement locally defined users and are investigating Lightweight Directory Access Protocol (LDAP) as a means to more securely provide this functionality. The new grid architecture also includes an NAS node which (like the Master node) is neither a submission nor execution node, serving instead as a centralized point of Network Attached Storage for home directories and user datasets NFS mounted onto grid submission and execution nodes, making user management, password administration, permissions, and quotas easier to administer as both grid and user base grow.

Investigations into grid architecture and implementation have revealed that although the technology is not yet mature, existing applications and protocols can be used to create a usable and maintainable grid environment with an acceptable level of customization. In the future, CBR will be investigating better data access methods for distributed databases (virtual data views and database federation) and interfacing the CBR grid with the cluster grid of the High Performance Virtual Computing Laboratory, a consortium of four universities in eastern Ontario, Canada, that was discussed earlier in the chapter. This linkage will be made using Globus to provide single sign-on portal access to all CBR resources, many of which will be configured as grid services. In this way, CBR hopes to provide enhanced

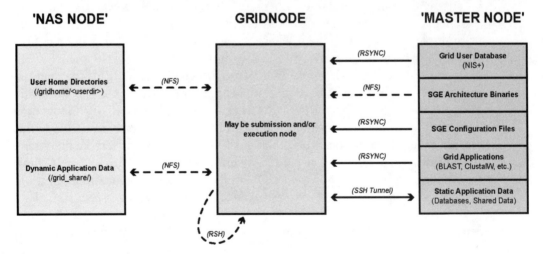

FIGURE 16.9   Revised NRC-CBR Global Grid architecture, including Network Attached Storage.

---

[42] Secure Shell, SSL, *http://www.openssh.com/*

bioinformatics services which more efficiently and intelligently use the computing resources distributed across Canada.

16.4.5 White Rose Grid

The White Rose Grid (WRG[43]), based in Yorkshire, UK, is a virtual organization comprising three universities: The universities of Leeds, York, and Sheffield. There are four significant compute resources (cluster grids) each named after a white rose. Two cluster grids are sited at Leeds (Maxima and Snowdon) and one each at York (Pascali) and Sheffield (Titania).

The White Rose Grid is heterogeneous in terms of underlying hardware and operating platform. Maxima, Pascali, and Titania are built from a combination of large symmetric memory Sun servers and storage/backup, and Snowdon comprises a Linux/Intel based compute cluster interconnected with Myricom Myrinet.

The software architecture can be viewed as four independent cluster grids interconnected through global grid middleware and accessible, optionally through a portal interface. All the grid middleware implemented at White Rose is available in open source form.

The WRG software stack, Figure 16.10, is composed largely of open source software. To provide a stable HPC platform for local users at each site Grid Engine Enterprise Edition,[2, 3] HPC ClusterTool[28] and SunONE Studio[44] provide DRM and MPI support and compile/debug capabilities.

Users at each campus use the Grid Engine interface (command line or GUI) to access their local resource. White Rose Grid users have the option of accessing the facility via the portal. The portal interface to the White Rose Grid has been created using the Grid Portal Development Kit[25] (GPDK) running on Apache Tomcat.[45] GPDK has been updated to work with Globus Toolkit 2.0 and also modified to integrate with various e-science applications.

Each of the four WRG cluster grids has an installation of Grid Engine Enterprise Edition. Globus Toolkit 2.0 provides the means to securely access each of the cluster grids through the portal.

---

[2] Sun Grid Engine Web site, *http://www.sun.com/grid*

[3] Grid Engine open source project at *http://gridengine.sunsource.net/*

[25] The Grid Portal Development Kit Web site is at *http://doesciencegrid.org/projects/GPDK/*

[28] The source for Sun HPC ClusterTools can be downloaded from *www.sun.com/solutions/hpc/communitysource*

[43] White Rose Grid, *http://www.informatics.leeds.ac.uk/pages/05_facilities/06_grid.htm*

[44] Sun ONE Studio, *http://wwws.sun.com/software/sundev/*

[45] Apache Tomcal server, *http://jakarta.apache.org/tomcat/*

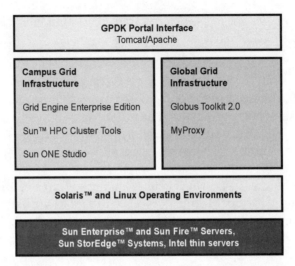

FIGURE 16.10 White Rose Grid components, hardware/software stack.

16.4.5.1 Grid Engine Enterprise Edition

Grid Engine Enterprise Edition (GEEE) is installed at each of the four nodes—Maxima, Snowdon, Titania, and Pascali. The command line and GUI of GEEE is the main access point to each node for local users. The Enterprise Edition version of Grid Engine provides policy driven resource management at the node level. There are four policy types which may be implemented:

- Share Tree Policy—GEEE keeps track of how much usage users/projects have already received. At each scheduling interval, the Scheduler adjusts all jobs' share of resources to ensure that users/groups and projects get very close to their allocated share of the system over the accumulation period.
- Functional Policy—Functional scheduling, sometimes called priority scheduling, is a non-feedback scheme (i.e., no account taken of past usage) for determining a job's importance by its association with the submitting user/project/department.
- Deadline Policy—Deadline scheduling ensures that a job is completed by a certain time by starting it soon enough and giving it enough resources to finish on time.
- Override Policy—Override scheduling allows the GEEE operator to dynamically adjust the relative importance of an individual job or of all the jobs associated with a user/department/project.

At White Rose, the Share Tree policy is used to manage the resource share allocation at each node, as shown in Figure 16.11. Users across the three universities

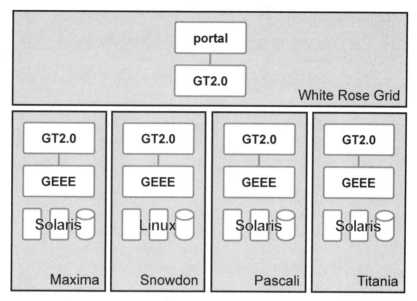

FIGURE 16.11 The four cluster grid computing nodes of the White Rose Grid—Portal Environment, Grid Service Provider, Grid Infrastructure, and Data Management System.

are of two types: (a) local users are those users who have access only to the local facility (b) WRG users are users who are allowed access to any node in the WRG. Each WRG node administrator has allocated 25 percent of their node's compute resource for WRG users. The remaining 75 percent share can be allocated as required across the local academic groups and departments. The WRG administrators also agree upon the half-life associated with GEEE so that past usage of the resources is taken into account consistently across the WRG.

16.4.5.2 Globus

As depicted in Figure 16.11, each WRG cluster grid hosts a Globus gatekeeper. The default job-manager for each of these gatekeepers is set to Grid Engine using the existing scripts in the GT2.0 distribution. In order for the Globus job manager to be able to submit jobs to the local DRM, it is simply necessary to ensure that the Globus gatekeeper server is registered as a submit host at the local Grid Engine master node. The Globus grid-security file referenced by the gatekeeper servers includes the names of all WRG users. New users' grid identities must be distributed across the grid in order for them to be successfully authenticated. Additionally to this, at each site all WRG users are added to the userset associated with the WRG share of the GEEE controlled resource. This ensures that the sum usage by WRG users at any cluster grid does not exceed 25 percent.

16.4.5.3 Portal Interface

The portal technology used at White Rose has been implemented using the Grid Portal Development Kit. GPDK has been designed as a Web interface to Globus. GPDK uses Java Server Pages (JSP) and Java Beans and runs in Apache Tomcat, the open source Web application server developed by Sun Microsystems. GPDK takes full advantage of the Java implementation of the Globus CoG toolkit.

GPDK Java Beans are responsible for the functionality of the portal and can be grouped into the five categories; Security, User Profiles, Job Submission, File Transfer, and Information Services. For security, GPDK integrates with MyProxy.[46] MyProxy enables the portal server to interact with the MyProxy server to obtain delegated credentials in order to authenticate on the user's behalf.

Some development work has been done in order to port the publicly available GPDK to GT2.0. Specifically, GPDK was modified to work with the updated MDS in GT2.0. Information Providers were written to enable Grid Engine Queue information to be passed to MDS. Grid users can query MDS to establish the state of the DRMs at each ClusterGrid.

As with many current portal projects, the WRG uses the MyProxy Toolkit as the basis for security. Prior to interacting with the WRG, a user must first securely pass a delegated credential to the portal server so that the portal can act upon that user's behalf subsequently. The MyProxy Toolkit enables this.

The event sequence up to job submission is as follows:

- When the user initially logs on, the MyProxy Toolkit is invoked so that the portal server can securely access a proxy credential for that user.
- The users can view the available resources and their dynamic properties via the portal. The Globus MDS pillar provides the GIIS, LDAP-based hierarchical database, which must be queried by the portal server.
- Once the user has determined the preferred resource, the job can be submitted. The job information is passed down to the selected cluster grid where the local Globus gatekeeper authenticates the users and passes the job information to Grid Engine Enterprise Edition.

16.4.6 Progress: Polish Research on Grid Environment on Sun Servers

The Poznan supercomputer project PROGRESS (Polish Research on Grid Environment for SUN Servers)[47] aims at building an access environment to computational services performed by a cluster of SUN systems. It involves two academic sites in Poland: Cracow and Poznan. The project was founded by the State Committee for

---

[46] Further information on MyProxy can be found at *http://www.ncsa.uiuc.edu/Divisions/ACES/MyProxy/*

[47] PROGRESS project home page, *http://progress.psnc.pl/*

Scientific Research. Project partners are: Poznan Supercomputing and Networking Center (PSNC);[48] Academic Supercomputing Center of University of Mining and Metallurgy, Cracow; Technical University Lodz; and Sun Microsystems Poland.

Currently, there are two cluster grids accessible through the PROGRESS portal: one Sun Fire 6800 (24 CPUs) in Cyfronet Cracow and two Sun Fire 6800 (24 CPUs) connected using Sun Fire Link sited at Poznan. The distance between the locations is about 400 km. Both locations also use Sun Fire V880 and Sun Storedge 3910 as the hardware supporting the Distributed Data Management System discussed below. At the development stage, only large Sun SMP machines are used, but the architecture allows the existing computing resources to be augmented by hardware from other vendors.

Globus Toolkit 2.0 (and 2.2) have been implemented to provide the middleware functionality. Sun Grid Engine Enterprise Edition is installed to control each of the cluster grids. The portal interface has been built using Web Services elements based upon J2EE.

The main task of this project is to give a unified access to distributed computing resources for the Polish scientific community. Other aims are:

- the development of novel tools supporting grid-portal architecture (grid service broker, security, migrating desktop, portal access to grid)
- the development and integration of data management and visualization modules enabling the grid-portal environment for other advanced applications (PIONIER program)

16.4.6.1 System Modules

The PROGRESS architecture can be described in terms of its constituent modules as shown in Figure 16.12. As well as using the Globus Toolkit, each of these modules provides a major piece of functionality for the PROGRESS grid. The four main modules are:

- Portal Environment
- Grid Service Provider
- Grid Infrastructure
- Data Management System

16.4.6.1.1 Portal Environment

The PROGRESS computing portal is a bioX thematic Web portal, which provides users with possibilities of accessing grid resources underlying the GSP Grid Service

---

[48] PSNC Poznan Supercomputing and Networking Center, *http://www.man.poznan.pl/research/index.html*

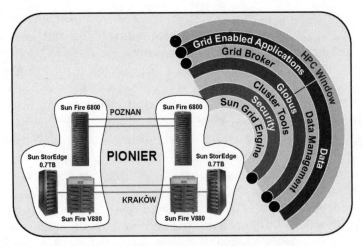

FIGURE 16.12 Overview of the PROGRSS Global Grid architecture.

Provider, applications collected in the PROGRESS application factory, and scientific data stored in the data management system. Additionally, the portal provides means of utilizing informational services of the GSP.

The PROGRESS GSP services are accessible through two client interfaces: WP (Web Portal) and MD (Migrating Desktop). The Web Portal, which is deployed on the Sun ONE Portal Server 7.0, performs three functions:

- grid job management—creating, building, submitting, monitoring execution, and analyzing results
- application management
- provider management

A screenshot depicting a typical user's view of the portal interface is shown in Figure 16.13.

The MD, which is a separate Java client application, provides a user interface for grid job management and DMS file system management. Both user interfaces are installed on a Sun Fire 280R machine, which serves as the PROGRESS system front-end.

Additionally, PROGRESS Portal gives access to services such as: news services, calendar server, and messaging server.

16.4.6.1.2 Grid Service Provider

The main module of the Progress access environment is the Grid Service Provider (GSP). It is a new layer introduced into the Grid Portal architecture by the

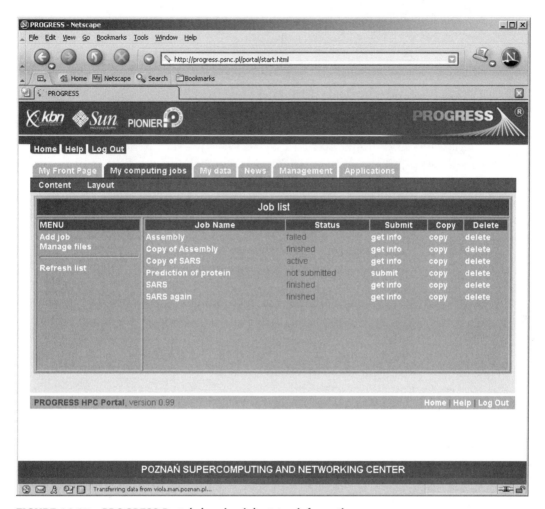

FIGURE 16.13 PROGRESS Portal showing job status information.

PROGRESS research team. The GSP provides users with three main services: a job submission service (JS), an application management service (AM), and a provider management service (PM):

- The JS is responsible for managing creation of user jobs, their submission to the grid, and monitoring of their execution.
- The AM provides functions for storing information about applications available for running in the grid. One of its main features is the possibility of assisting application developers in adding new applications to the application factory.

- The PM allows the GSP administrator to keep up-to-date information on the services available within the provider.

16.4.6.1.3 Grid Infrastructure

The Grid Resource Broker (GRB) is developed at PSNC and enables execution of PROGRESS grid jobs in the grid. Cluster grids are managed by Sun Grid Engine Enterprise Edition software with Globus deployed upon it. The GRB provides two interfaces: a CORBA interface and a Web Services interface. Grid job definitions are passed to the GRB in form of an XRSL document and the GRB informs the JS about events connected with the execution of a job (start, failure, or success).

Sun One Grid Engine Enterprise Edition is implemented at both sites as the local distributed resource manager. Grid Engine Enterprise Edition provides policy driven resource management at the node level. The Share Tree policy is used to manage the resource share allocation at each node.

Two types of users can have access to resources:

- local users, accessing compute resources through Grid Engine GUI
- portal users, accessing nodes using PROGRESS Portal or Migrating Desktop

In the event of extensions to the PROGRESS grid, Grid Engine would also be used. Each cluster grid also hosts Globus gatekeeper.

16.4.6.1.4 Data Management System (DMS)

PROGRESS grid jobs use the DMS to store the input and output files. The DMS provides a Web Services-based data broker, which handles all requests. The DMS is equipped with three data containers: the file system, the database system, and the tape storage system. A data file is referenced within the DMS with a universal object identifier, which allows for obtaining information on the location of the file. Users can download or upload file using one of three possible protocols: FTP, GASS, or GridFTP.

16.5 CONCLUSION

Grid deployments in research and academic institutions are serving as useful test beds as Grid Computing transitions to commercial enterprises. Grid deployments are using unique combinations of commercial and public domain open source software to build extremely powerful grids, which are solving problems in a wide array of disciplines. Early indications from industry grid deployments are clearly pointing to tremendous enterprise productivity and infrastructure performance gains.

Acknowledgments

The information in this chapter has been put together by many busy people. They still found time to write a summary about their grid implementations. Especially, I would like to thank:

- Suzanne George and Chris Nicholas, GlobeXplorer
- Barbara Chapman and Babu Sunderam, University of Houston
- Kenneth Edgecomb, HPCVL Ontario, Canada
- Jason Burgess and Terry Dalton, Canada NRC-CBR
- Cezary Mazurek and Krzysztof Kurowski from Poznan Supercomputing and Networking Center
- Ian Foster and Carl Kesselman from the Globus Project, and James Coomer, Charu Chaubal, and Radoslaw Rafinski from Sun Microsystems

17 Grids in Life Sciences

Matt Oberdorfer and Jim Gutowski

Engineered Intelligence

In This Chapter

- Life Sciences and Bioinformatics
- Impact of Grid Computing Technology on Life Sciences

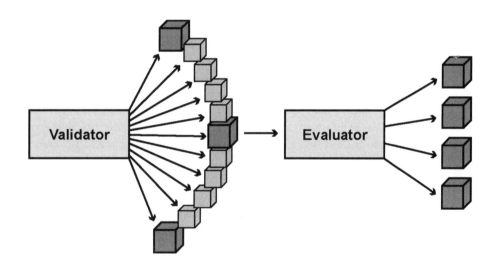

17.1 INTRODUCTION

Computer technology is an integral part of fundamental change underway in life sciences. *Biotechnology* captures the segment of life sciences that combines biology, chemistry, physics, mathematics, and computer science, and applies biological techniques developed through basic research-to-research and product development. Similar to the way computer technology changed product development in the automotive industry—where simulations run on parallel computers have essentially replaced physical prototypes for car crash studies—biotechnology uses parallel computing to dramatically improve understanding the building blocks of life—genes and proteins—and to enable scientists to test, categorize, and simulate the effects of new drugs on the human body. The recent identification and publication of the human genome has pushed scientists to adopt new computing technology even more. (A *genome* is the total genetic information possessed by an individual organism. Each cell contains a complete copy of the genome.)

Grid Computing plays an essential role in two key areas of biotechnology: in bioinformatics, which focuses on the search for information and comparison of new information to that which is already established (such as with the human genome, which requires extensive database technology); and in computational chemistry and biochemistry, which focus on development of new chemical structures using techniques—such as molecular modeling—that require high-performance computing technology.

Biotechnology companies have been turning to high-performance computers to help manage the weight of data generated by modern biology. Modern approaches to biology, such as genomics, which looks for individual genes in the variety of human DNA, or proteomics, which tries to describe some of the most complicated molecules in the body, require sifting through massive amounts of data. Scientists deal with basic elements, genes, and proteins, and work to understand their structures—which define their functions—and the interaction between chemical compounds and these complex structures. Using parallel computing, they can compare structures to those already defined, such as in the Human Genome Project, and simulate new ones with computation biology and chemistry.

In dealing with these challenges, life sciences companies are redefining their research methodologies and retooling their IT infrastructures to position themselves for success in this new environment. The traditional trial-and-error approach is rapidly giving way to a more predictive science based on sophisticated laboratory automation and computer simulation. Computing technology used in newly emerging life sciences discovery models is critical to researcher productivity and time to market. As new approaches emerge—such as Grid Computing and desktop supercomputing—more scientists are able to tackle increasingly complex problems.

17.2 BIOINFORMATICS

Grid Computing is becoming more prevalent in both basic research and drug development—the industry that takes biotech from research to commercial applications. Computer technology is used at nearly every stage of the drug development process. In the early stages of pre-clinical testing, research, and development, bioinformatics technology allows researchers to analyze terabytes of data being produced by the Human Genome Project. Bioinformatics is the discipline of obtaining information about genomic or protein sequence data. This may involve similarity searches of databases, comparing an unidentified sequence to the sequences in a database, or making predictions about the sequence based on current knowledge of similar sequences. Gene sequence databases, gene expression databases (which track how genes react to various stimuli), protein sequence databases, and related analysis tools all help scientists determine whether and how a particular molecule is directly involved in a disease process. That, in turn, aids in the discovery of new and better drug targets.

Bioinformatics may be the only way drug companies can deal with the gigabytes of data they produce and receive every day. The sequenced human genome has already increased the number of biological drug targets that can be explored from about 500 to more than 30,000—and that number will only get larger. The completion of a working draft of the human genome—an important milestone in the Human Genome Project—was announced in June 2000 and published in February 2001. Databases available via the Internet put this wealth of information at the disposal of the world's scientific community. At the same time, computational chemistry allows companies to synthesize more than 100 compounds per chemist per year, and advances in computing technology continue to raise that number even higher.

17.2.1 Bioinformatics and High-Performance Computing

Bioinformatics in the life sciences industry is focused on the compilation of genomic information into databases and identifying and analyzing patterns within genetic data for viability as pharmaceutical products. Drug discovery and biotech research companies need powerful, high-performance solutions to search through and analyze huge amounts of data. High-speed, high-performance computing power and industrial-strength databases perform a wide range of data-intensive computing functions: mapping genetic and proteomic information, data mining to identify patterns and similarities, and text mining using huge libraries of information. These activities require high-performance computer infrastructures with access to huge databases of information.

These databases are massive and, as a result, researchers must compare each sequence with a vast number of other sequences efficiently. A number of programs have been written to rapidly search a database for a query sequence. Comparison, whether of structural features or protein sequences, lies in the heart of bioinformatics. The introduction of BLAST, or the Basic Local Alignment Search Tool, in 1990 made it easier to rapidly scan huge databases for overt homologies, or sequence similarity, and to statistically evaluate the resulting matches. BLAST works by comparing a user's unknown sequence against the database of all known sequences to determine likely matches. Sequence similarities found by BLAST have been critical in several gene discoveries.

17.2.2 Grid Computing and Bioinformatics

Database searching and comparison or sequencing is embarrassingly parallel with the application able to be split up and run independently—there is no need for communication between the processors doing the comparison, simply the sharing of results when complete. Grid Computing is emerging as one way to make this search happen fast, using untapped computing resources in the company or, in some cases, around the world via the Internet to run these sequences. Each computer in the grid (the collection of all the computers connected by a network) is given a small subset of the main task, such as comparing a particular protein structure to part of a large database, and carries it out independent of any other system in the grid.

Most major pharmaceutical companies have grid-computing projects underway to explore the possibilities in this fast-growing area. A typical project at a pharmaceutical company runs software that analyzes the efficacy of potential drug compounds using the spare computing cycles of several thousand PCs to speed drugs to market. When PCs are not otherwise being used, Grid Computing harnesses spare computing power to run data comparison in parallel. This approach works well for bioinformatics applications, but does not fit when interaction between the processors is required, such as in molecular modeling.

The considerable "algorithmic complexity" of biological systems requires a large amount of detailed information for their complete description. There has been a vast quantity of information gathered for all kinds of biological systems at the molecular and cellular level that now require computational tools to be adequately interpreted.

BOX 17.1 Bioinformatics

Bioinformatics—An Excellent Grid Application

A field of science in which biology, computer science, and information technology merge to form a single discipline. The ultimate goal of the field is to enable the discovery of new biological insights as well as to create a global perspective from which unifying principles in biology can be discerned. At the beginning of the *genomic revolution*, a bioinformatics concern was the creation and maintenance of a database to store biological information such as nucleotide and amino acid sequences. Development of this type of database involved not only design issues, but the development of complex interfaces whereby researchers could access existing data as well as submit new or revised data.

Because of their tremendous computational requirement, it is not a surprise that life sciences companies are early adopters of Grid Computing technology. Another equally important fact is that life sciences companies have substantial experience in distributed computing based on clusters and high-performance computing. Cluster and distributed computing have been a core component of the IT infrastructure of many of the large- and medium-sized life sciences companies. Penetration rates are 85 percent in large pharmaceutical companies and 65 percent in small- to medium-sized life science organizations.[1]

Applications developed for analysis of genome sequencing used in the Human Genome Project, and optimized over the years, have been ported over by Grid Computing vendors and customers alike for use on desktop grids. One such application, BLAST, is a set of similarity search programs designed to explore all of the available sequence databases regardless of whether the query is protein or DNA.[2] United Devices offers a version of the application optimized for its Metaprocessor platform. IBM has partnered with TurboGenomics, the creator of TurboBLAST, to create a product called IBM-TurboBLAST. This is a bundled product that includes the application optimized to run on an IBM eServer pSeries 690 or an IBM eServer Cluster 1300 that runs the Linux operating system. Blackstone Computing's Powercloud application has also optimized BLAST to utilize clusters and desktop grids.

---

[1] InSilico Research report and Grid Technology Partners industry interviews.

[2] The Basic Local Alignment Search Tool (BLAST) has been developed by the National Center for Biotechnology Information, a National Institute of Health funded organization, *www.ncbi.nlm.nih.gov*

Another application, DOCK, addresses the problem of *docking* molecules to each other. It explores ways in which two molecules, such as a drug and an enzyme or protein receptor, might fit together. Compounds which dock to each other well have the potential to bind. A compound which binds to a biological macromolecule may inhibit its function and act as a drug; thus making DOCK one of the most critical tool in proteomics research today.

Life sciences applications such as HMMER, CHARMm, LigandFit, GOLD, Delphi, and Archimedes also exhibit good parallel efficiency and are being deployed on grids today.

17.2.3 Example Grid Computing: Smallpox Research

The Smallpox Research Grid Project, being driven by the United States Department of Defense, shows the potential to harness spare computing power across the country for research. The project will be powered by a massive computing grid that will enable millions of computer owners worldwide to contribute idle computing resources to the task of developing a wide collection of potential anti-smallpox drugs.

The project is based on commercially available technologies and services used by many pharmaceutical and biotechnology companies to improve and accelerate drug discovery and development. The project will employ computational chemistry to analyze chemical interactions between a library of 35 million potential drug molecules and several protein targets on the smallpox virus in the search for an effective antiviral drug to treat smallpox post-infection. Past projects, such as one for anthrax, have been able to complete the billions of computer simulations in as little as a month.

17.3 COMPUTATIONAL CHEMISTRY AND BIOCHEMISTRY

On the other side of the development process, there is complex modeling and simulation to develop new compounds. Drug companies employ a variety of computational chemistry and biochemistry software—tools that can predict the activity of a particular compound by studying its molecular structure. For instance, scientists can use molecular modeling software, with interactive 3-D visualization or mathematical algorithms, to discover and design safe and effective compounds. As in bioinformatics, chemical databases allow researchers to store and retrieve compounds and related data.

17.4 PROTEIN MODELING

DNA sequences encode proteins with specific functions. Traditionally, a protein's structure was determined with x-ray crystallography or NMR (nuclear magnetic resonance) spectroscopy. Researchers have been working for decades to develop procedures for predicting protein structure that are not as time consuming and are not hindered by size and solubility constraints. To do this, researchers have turned to computers for help in predicting protein structure from gene sequences. Using available genomic information, such as with the Human Genome Project, this goal can be approached in a logical and organized fashion, and researchers can attempt to predict the three-dimensional structure using protein or molecular modeling. This method uses experimentally determined protein structures to predict the structure of another protein that has a similar amino acid sequence.

Although molecular modeling may not be as accurate in determining a protein's structure as experimental methods, it is still extremely helpful in proposing and testing various biological hypotheses. Molecular modeling also provides a starting point for researchers wishing to confirm a structure through x-ray crystallography and NMR spectroscopy. As the different genome projects are producing more sequences, and because novel protein folds and families are being determined, protein modeling will become an increasingly important tool for scientists working to understand normal and disease-related processes in living organisms.

Identifying a protein's shape, or structure, is key to understanding its biological function and its role in health and disease. Illuminating a protein's structure also paves the way for the development of new agents and devices to treat a disease. Yet solving the structure of a protein is no easy feat—it often takes scientists working in the laboratory for months, sometimes years, to experimentally determine a single structure. Therefore, scientists have begun to utilize computers to help predict the structure of a protein, based on its sequence. The challenge lies in developing methods for accurately and reliably understanding this intricate relationship.

17.5 *AB INITIO* MOLECULAR MODELING

The term *ab initio* is Latin for "from the beginning." This name is given to computations derived directly from theoretical principles with no inclusion of experimental data; usually, this refers to an approximate quantum mechanical calculation. The approximations made are usually mathematical approximations, such as using a simpler functional form for a function or getting an approximate solution to a differential equation.

The positive side of *ab initio* methods is that they eventually converge to the exact solution, once all of the approximations are made sufficiently small in magnitude. The negative side is that they are computationally expensive. *Ab initio* methods often take enormous amounts of computer CPU time, memory, and disk space. The most common method scales as N^4, where N is the number of basis functions, so a calculation twice as large takes 16 times as long to complete. Correlated calculations often scale much worse than this. In general, *ab initio* calculations give very good qualitative results and can give increasingly accurate quantitative results as the molecules in question become smaller.

17.5.1 Example *Ab Initio* Modeling with Parallel Computing

In 2002, a research project in the department of chemistry and biochemistry at a major university confirmed the power of parallel computing for *ab initio* molecular modeling. A small portion of a large *ab initio* FORTRAN program, which had run serially on a single computer system, was added in CxC parallel language to parallelize the most complex matrix mathematics of the application. Running this parallel code on a cluster grid of 20 computers realized a 63X performance improvement. Molecular modeling can require calculation using massive matrixes; in this case, the total matrix size was more than 10^{12}, requiring the research team to split up the problem and compute sub-matrixes to get a final result. With this type of approach, researchers can tackle larger molecules and more complex problems.

17.6 GRID COMPUTING IN LIFE SCIENCES

The variety and range of computing uses in biotechnology and drug development present a number of opportunities for both grid and high-performance computing. The massive amounts of data that need to be analyzed and the complexities of molecular modeling both point to parallel computing as a means to shorten the time to result.

Grid Computing is emerging as a promising approach to speed up analyses for database searches, data mining, and sequencing. It divides tasks into parallel programs that execute independently, even using spare computing cycles to reduce cost. For applications such as molecular modeling, it is imperative that the processors doing the calculations interact—calculations of interaction between atoms are inherently interconnected and require high-speed data transfer to generate results. These problems cannot be run on grids, but instead require supercomputing using high-speed data transfers, either through a local memory bus in a shared-memory supercomputer or through high-speed networking in supercomputing clusters.

Parallel computing is a key ingredient in the development of new algorithms with which to assess relationships among members of large data sets, such as methods to locate a gene within a sequence, predict protein structure and/or function, and cluster protein sequences into families of related sequences.

17.7 ARTIFICIAL INTELLIGENCE AND LIFE SCIENCES

Techniques invented by computer scientists interested in artificial intelligence have been applied to drug design in recent years. The general scenario is for a target to be identified, then developing a structure for a molecule to interact with this target in order to change its functionality is desired. Rather than have a chemist try hundreds or thousands of possibilities with a molecular-modeling program, molecular modeling is built into an artificial intelligence program, which tries enormous numbers of reasonable possibilities in an automated fashion.

Genetic algorithms (GA) can gather the best results and feed them into further solutions. The algorithm starts with a set of solutions called a *population*. After simulation, solutions from one population are taken and used to form a new population. This is motivated by a hope that the new population will be better than the old one. Solutions that are selected to form new solutions (*offspring*) are selected according to their fitness—the more suitable they are, the more chances they have to reproduce. Using GA routines, scientists can evolve their structures to more efficiently find and develop those best suited to the particular task.

17.8 CONCLUSION

Emergence of new approaches in bioinformatics and developing computer technology go hand-in-hand in redefining the shape of life sciences for the future. Analyses of the draft human genomic sequence have already led to the identification of genes for cystic fibrosis, breast cancer, hereditary deafness, hereditary skeletal disorders, and a form of diabetes—just to name a few. The draft sequence has also been used to identify an enormous number of variations in the genetic code that play a significant role in disease processes. These discoveries, as well as future discoveries, will have a profound impact on the future conduct of biomedical and pharmaceutical research.

In high-performance computing, development of grid technologies and improvements in parallel computing with clusters and desktop supercomputing make the power needed for these complex tasks available to every scientist and research engineer. Faster data mining across huge databases, coupled with parallel processing and improved modeling solutions, will bring about radical improvements to drug discovery and development processes.

18 Grids in the Telecommunications Sector

Dr. Marco Laucelli

GridSystems

Prof. Joan Massó

GridSystems

In This Chapter

- Grid Computing for Enterprise Telecommunication Applications
- Grid Computing as a Telecom Service

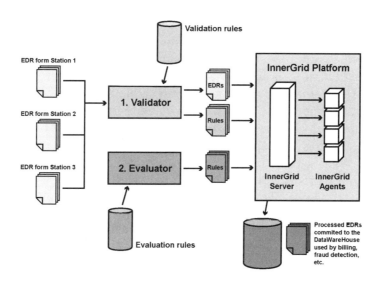

18.1 INTRODUCTION

The chapter focuses on the application of Grid Computing in the telecommunications sector. For the sake of brevity, we will use the shortcut term *Telco* to refer to this sector.

We will study two experiences of GridSystems' implementing grid technology in Telco environments. One refers to high CPU-demanding network planning systems, and the other explains how to build a grid-based DataWareHouse for Enterprise Data Records (EDR) analysis. These two experiences will motivate the proposal of corporate grid platforms in Telco companies. In the second part of this chapter we will focus on a slightly more speculative topic. We will describe how the grid network could work in the next years and the potential role of Telcos. In that scenario, Telcos will be considered not just as users of the technology, but especially as providers. They will not only offer network services, but also they could offer their grid knowledge to their customers.

18.2 TELCOS AS USERS

In many of the scientific grid projects that have been performed during the last years, few processes can be considered as mission critical, and many demonstrations have been shown to work in a best case or one shot basis. In a commercial environment, companies do have well-defined periodical time windows to complete their jobs. In these cases, peak performance is a critical requirement. In this section we will illustrate two uses of grid technology that match the performance requirement of business processes warranting scalability and return on investment.

During the past few years, corporations have heavily invested in IT, and even entered into the realm of what has been traditionally called HPC (high-performance computing). Unfortunately, HPC is extremely dependent on expensive hardware, implying high costs of ownership. Corporations have in their networks a large number of heterogeneous high-end servers, which are used to warrant performance in critical processes, but they are underused during large periods. One way to reduce the cost of HPC would be to join resources in a computing network. This is the idea behind old fashioned metacomputing,[1] which involves high costs of administration, and the lack of fully robust, reliable, and resilient solutions. These two problems of metacomputing technologies compromise the scalability of processes,

---

[1] Foster, C. Kesselman. "Globus: A Metacomputing Infrastructure Toolkit," *Intl J. Supercomputer Applications*, 11(2):115–128, 1997.

even when the corporation owns enough computational power. Another possible way to reduce costs could be through the use of some High Throughput Computing[2] systems to scale processes using any kind of underutilized computational resource. Most of these systems were born in the research community and their current state of commercial support and lack of transparent easy-to-use solutions makes them unsuitable for most corporate computational processes.

The most effective and efficient solution comes from what we call Enterprise High Throughput Grids (EHTG), which allow an easy and robust integration of the whole corporate network in a computing platform. Companies implanting an EHTG can transform their sparse and heterogeneous computers—high-end servers, workstations, desktop PCs—in a single virtual utility. EHTG also allows establishing collaborations among departments by the definition of execution policies to share their resources. In critical periods one department could require more computational power than it owns. The department could ask the EHTG to find and use underused computational resources of other departments. All these features make Enterprise High Throughput Grids the only viable technology available today to scale business processes.

We will give some examples of the use of EHTG in the telecommunications sector. Both experiences have been carried out using *InnerGrid*, the grid technology of *GridSystems*.

18.2.1 CPU Intensive Application: Network Planning and Management

Network planning and network management are two key processes that determine the competitiveness of a telecommunications operator. Long-term network planning determines the strategic and technological position of a telecommunications operator. For example, the planning of future 3G networks took off some years ago, allowing the Telcos to have a clear position in the future market. Short-term planning determines the available services and costs and hence the economical benefits of Telcos. In this sense short-term planning is closely linked to network management processes. Telecommunications operators require short-term flexible planning systems that allow them to plan networks at always shorter times so they can have more efficient and rapid responses to customer demands as well as market conditions.

The conditioning factors of network configurations have increased their complexity due to new type of services (such as voice services, SMS, TCP/IP, etc.) and they will suffer an even larger growth with the implantation of new generation services. For these reasons, network planning and management processes have also increased their complexity. On one hand, the algorithms used for network planning

---

[2] The Condor Project, *http://www.cs.wisc.edu/condor/*

have had to integrate the ability to analyze this new situation. This is the reason artificial intelligence algorithms have been implanted (as neural networks, genetic algorithms, simulated annealing, etc.) On the other hand, network traffic data to be analyzed has increased considerably. From a computational point of view, the main consequence of these increasing requirements is the need for larger and larger computational power. In the future, this tendency will become even more pronounced. The computational requirements of new network planning systems and the need of planning at always shorter terms require a solution that can warrant the scalability of the process at acceptable costs. As argued above, grid technology in general and nowadays EHTGs are the only possible solution.

Let us consider a concrete example of short-term network planning, related to a network management system. The system uses as input network traffic data obtained from processing of EDRs (see below) and starts a simulated annealing optimization based on the simulation of future scenarios of network traffic. After the iterated annealing process, a new planning of the network is obtained and linked to the management processes. The simulation of scenarios could be processed in independent pieces distributed by the grid server among the nodes of a grid platform. Each node processes a single scenario and gives the results back to a centralized set of federated computers. The simulated annealing algorithm produces a new test solution for the network planning, and a new iteration of the procedure is started. The process is iterated until convergence to an optimum solution is achieved (see Fig. 18.1).

This grid approach to solve short-term planning computational requirements has given excellent results. Each scenario could be processed in a desktop PC, so, at least for a Telco company, the number of nodes that could be included in the EHTG platform for this process is almost unlimited. The amount of data required for the simulation of one scenario is not very large, so the communication time spent passing these data to each node is not compromising. As a result, one finds that the solution scales very well, allowing the use of a very large amount of non-dedicated computers for the network planning processes.

In addition to the performance of the system, there are two business requirements that the grid solution should satisfy:

- *Complex workflow*: The implementation of an iterated process in a distributed platform requires the possibility of defining complex workflows. The system should interact with the grid platform in a dynamic way. The distributing procedure should be integrated in the logical workflow of the process. In the case of *InnerGrid* this has been very simple, using its API to define the interaction between the Simulated Annealing Manager and the grid platform. Other grid platforms would require significantly more programming from the part of the solution developer.

- *Criticality*: When linked to a network management system, the network planning system should warrant the execution of processes in time window. Telcos are used to run this type of applications on supercomputers with a precise measure of the execution time. Going to a distributed and heterogeneous platform could imply losing the estimation of the required processing time. This must be avoided by the system. An EHTG such as *InnerGrid* allows the user to define an accurate execution policy to warrant the completion of the analysis in a user-defined time window.
- *Fault tolerance*: Related to the previous point, the system should be fully fault-tolerant. This requires that all the elements of the system must have checkpoint-restart mechanisms and there must be no single point of failure in the system (e.g., the central server of the grid platform must be associated to redundant servers that will warrant a full-time availability of the services of the platform). In this sense, *InnerGrid* does provide fully robust, reliable, and resilient EHTG capabilities.

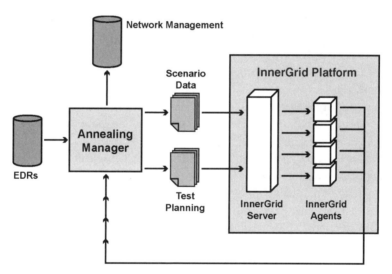

FIGURE 18.1 An iteration of a simulated annealing process for short-term network planning.

The previous example shows that an EHTG matching all these requirements and allowing a scalable performance is a viable solution for a high CPU-demanding application. In the concrete case of a network planning system linked to network management processes, grid technology has shown its capability to shorten the planning time, therefore giving the telecommunications operators the opportunity to be much more agile and efficient on this aspect of their business, without facing high costs in computational resources.

More examples of high CPU-demanding applications can be found in the Telcos processes. In the case of long-term network planning, it is possible to distribute artificial intelligence algorithms to plan complex network configurations. Other examples can be found in CRM processes, quality of services, etc.

18.2.2 Data Intensive Applications: EDR Analysis and DataWareHouse

One of the most data intensive processes that must be carried out by a telecommunication operator is the analysis of data regarding the use of the network services that it provides. The analysis of these data is necessary for almost all the activities in the Telcos. Extracting information from these data is relevant, for example, to knowing the behavior of customers and launching adequate marketing strategies. The same data are needed to plan the network, to prevent fraud, to evaluate the quality of the service, or to execute the billing process.

Nowadays, the registered data are principally related to voice and SMS services used by customers. In the near future the situation will change. Telcos are going to offer an ample variety of 3G services that will require the analysis of many more data and would allow knowing many more details on customers' behavior and network. This knowledge will be crucial to expand the business of the companies in the new multimedia market.

Data related to calls are stored in CDRs (Call Data Records) files. These files are processed continuously in high-end machines and then stored in a DataWareHouse that is used by several business processes. A 6-million-user operator may process about 200 million CDRs a day. The processing time of these files almost reaches the capacity of a current supercomputer. In the next few years, registered data will include information about many other activities carried out by users, in addition to voice services. Already today, CDRs have been extended to EDRs to include, for instance, SMS services. With the arrival of the new 3G services, the information contained in EDRs will increase. As a consequence, the computational needs to process these data and convert it into useful business information will compromise traditional hardware-based solutions. Once again, grid technology is seen as the only viable affordable solution.

Let us analyze an example in which an EHTG has been used to carry out the analysis of EDRs. It is a two-step process. The first step consists of a validation of the data contained in the EDRs and arriving from the network stations. The system loads the validation rules from a database, and sends them to the grid platform. The central grid server distributes the EDRs among the nodes for their process in a distributed mode.

Once the validation step has been completed, an evaluation of the EDRs takes place. In this evaluation, the EDRs data are transformed for its inclusion in the DataWareHouse. The system loads the evaluation rules from a database, and sends them to the platform. The evaluation of the validated EDRs is distributed among the computers in the platform. Finally the results are committed to the DataWare-House (see Fig. 18.2).

The differences between the *validation* and *evaluation* processes produce some interesting features of the system. The validation process needs much less computational power than the evaluation process. Therefore it could be executed in less powerful computers, any kind of desktop PC, for example. The evaluation process, however, requires some minimal power. It needs to be executed in workstations or at least in powerful PCs. This difference between step one and two is the major reason to split the whole procedure in two. Doing this, one could define execution policies in the grid platform to allow the distribution of the validation process among many more computers than the evaluation process. Having split the process in two and using correct execution policies allows profiting from many more computers than if a single process had been defined (Figure 18.3).

There are two main lessons we may learn from this experience:

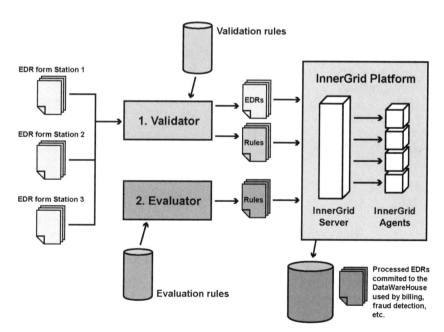

FIGURE 18.2 A two-step EDR processing platform, using InnerGrid.

■ *Execution policies*: a grid platform should be able to use precise execution policies. This requirement should provide the platform with the ability to discover and use the best computational resources for each job. By *best computational resource*, we mean the cheapest available resource matching the requirement of the job. Execution policies are closely linked to workflow functionalities. In the previous example, it would have been impossible to define a two-step execution—and the corresponding execution policies—without a definition of the logical workflow. With both functionalities, it is possible to make an accurate use of the resources in the grid platform, giving the corporation the maximum profit from its internal computational power. Grid technology solutions should provide these services in an easy-to-use way, as *InnerGrid* does, without requiring expertise or technical efforts from the users.

■ *Virtualization and scalability*: We have seen that there could be processes to be run in a grid platform that would use a different type and number of resources during different steps. As a consequence, when measuring the scalability of the solution, we cannot use the traditional method of comparing execution time to the number of nodes included in the platform. In fact, in this case it is not even possible to define a number of nodes for the process.

The underlying idea here is the *virtualization*. The internal structure of the grid platform is something unknown to the users, so they should use other criteria to measure the performance and the scalability of the utility.

In addition to the execution policies and scalability questions analysed here, the arguments about fault tolerance and criticality given for the network planning case apply here directly.

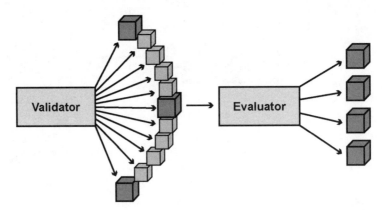

FIGURE 18.3 InnerGrid Resource Broker finds adequate computers for each step.

Before proceeding, let us mention some results obtained in the EDR analysis grid platform. Scalability, understood as the response of the platform to changes in the task size (the number of incoming EDR files), has been completely achieved. With a relevant number of computers in the platform, the system has enough degrees of freedom to statistically find adequate available resources, warranting the execution of the task in a time window. Because these kinds of processes require high data communications, it is necessary to tune the distribution process to find the optimum equilibrium between communication and distribution. This has been done allowing changes of the size of the computational task assigned to each node. Tuning this parameter one finds an optimum scalability of the system.

Similar examples of using grid technology in data-intensive applications could be found in other Telcos processes. As we will see, it is possible to use a grid platform in fraud detections, billing processes, or quality of service analyses.

18.2.3 A Corporate Grid Platform

The analysis of the use of an EHTG platform allows us to define the requirements to be achieved by the technology to be used in Telcos' business processes. These requirements are: support for heterogeneous platforms, full fault-tolerance, workflow definition functionalities, execution policies definition to warrant maximum profit form computational power, and time windows definition for critical processes. If the grid technology offers the suitable functionalities to match these requirements, without requiring extra efforts from the users, then a corporation could consider building a grid platform to convert its computational resources in a single virtual utility. These kinds of utilities have been sometimes called *internal data center*.

Let us consider a Telco operator with several processes linked to the analysis of EDRs. Each process needs high computational power in critical periods (for example, the end of the month for the billing process) and has dedicated computers (Fig. 18.4). It is possible to implement all these processes in a grid platform by adding non-dedicated computational power to dedicated resources. Apart from increasing the resulting computational capacity, the grid platform will allow the company to provide critical processes to extra computational power. For example, in the end-of-the-month billing period, some network planning resources could be moved to the billing applications while maintaining the EDR analysis and the fraud detection performance (Figure 18.5).

This example shows how a company could benefit from implanting a corporate grid platform. It allows distributing the computational resources of the company among the processes it is carrying out, depending on the performance required in critical periods. Grid technology, available today, allows the scaling of the processes of the Telcos companies to the challenges they must confront in the era of 3G wireless.

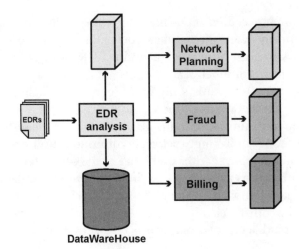

FIGURE 18.4 Business computational processes with dedicated computational resources.

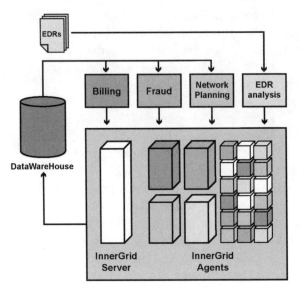

FIGURE 18.5 The internal grid platform in a critical billing period.

18.3 TELCOS AS PROVIDERS

Many of the discussions of next generation grid technology position it as an evolution of Web technology, coming from the fusion of Grid Computing and Web Services to produce a new Internet based on *Grid Services*.[3, 4] Grid Services will create an abstract environment to allow the sharing of any kind of hardware (computers, storage and communication devices, networks, mobile devices, etc.) and software (applications, scheduling services, etc.) resources. The possibility of joining and sharing resources will facilitate the collaboration of geographically distributed communities and will allow corporations to offer and request any kind of computational process. In short, Grid Services will create a new digital society that will use the new global network as a virtual computational utility. Grid Services will rely on four main pillars: the network, the standards to deploy services, the software to use and to manage utilities, and the knowledge to allow a profitable and transparent relationship between users and utilities.

At first, one might think that the primary role of Telcos in a global grid network would be to keep providing network bandwidth. A global grid network will offer Telcos the opportunity to sell their knowledge of the network infrastructure on which Grid Services will be deployed to companies that want to use the grid services without dealing with the underlying complexity.

18.3.1 Grid Services: A Business in the Future Network

Suppose that an ASP company is offering its computational resources, which have been set as a grid platform, to run Finite Element Analysis. It has deployed this capacity as a Web Service describing the type of analysis and the computational power available. Let us consider also a chemical company that has deployed a Web Service offering a database describing the chemical and physical properties of an ample variety of materials. Both these cases could be seen as examples of Grid Services. Now let us imagine an automobile company that wants to run the simulation of a new chassis design, using the services available on the network. They would like to simulate different materials, using the data provided by the chemical company and the Finite Element Analysis capacities provided by the ASP. The company should develop a system that uses the Grid Services to complete the simulations.

Let us also suppose that there are several companies on the Internet offering similar Grid Services, with different capacities and different functionalities. The

---

[3] *Next Generation Grid(s), European Grid Research 2005–2010.* Expert Group Report, June 16, 2003.

[4] John Hagel III and John Seely Brown "Service Grids: The Missing Link in Web Services," *http://www.johnhagel.com/*

company must search the Internet for all the Grid Services that could be suitable for their purpose, evaluate the quality and costs of each offer, and select the most suitable. Of course, there should not be any doubt about the quality of the services provided and the full security in the whole process. Secondly the selected service should achieve the execution time requirements defined by the criticality of the process. If the simulation should be carried out in one day, for example, the company should receive a guarantee that the result of the simulation will come in this time. And finally, the company should look for the cheapest service. With all these efforts it could mean that, for the automobile company, the model of running the simulations using Grid Services turns out not to be economically viable. And hence they might prefer to spend their money on internal computational resources, software licenses, and databases. This brings into play the role of a Telco.

The knowledge of the network could put the Telcos in a good position to offer companies, such as the automobile company, the opportunity to use the Grid Services without worrying about any technical detail. The automobile company will submit to the Telco a description of the simulation they want to carry out. This description will include all the business requirements for the process, including time windows and, for example, maximum costs. The Telcos will search among their customers to find the most suitable Grid Services and will manage the correct execution of the processes requested to them, giving back to the automobile company the results of the simulations. In the business literature, the model for the network that we have described is usually seen as an extension of the outsourcing model. Some have called it *e-sourcing*. In this framework, any kind of Grid Services will be available on the network, but there would be a need for companies to bring these utilities to the customers.[5] This would be the role of *aggregators*, and Telcos would be good candidates.

Grid Services would allow using almost any kind of device capacities through the network, so the type of processes required by customers could be as varied as we could imagine. Telcos should be able to guarantee a secure use of these new services on the Internet; they should also create the methods to discover and use Grid Services, and define accounting, billing, and quality of services models. A lot of work has to be done to create adequate standards; a completely defined business model for these activities on the Grid Network is still lacking. But the ideas are drawing nearer and the opportunities are coming with them.

18.4 CONCLUSION

In this chapter, a relationship between the telecommunications sector and grid technology was discussed. The experiences that have been detailed have shown that

---

[5] John Hagel III and John Seely Brown, *"Service Grids: The Missing Link in Web Services,"* http://www.johnhegel.com/

grid technology, and particularly EHTG platforms, are already available and profitable today. Telcos could build computational utilities with the computational resources that they already own. The emerging properties of these new platforms will allow the scalability of the Telcos' business processes that would suffer drastic changes with the next generation era. These changes will boost the required computational power, and only grid technology will meet the challenge at reasonable costs.

After analyzing the possible use of grid technology by Telcos, we imagined a Grid World where Grid Services would allow the use of any kind of device capacities through the network. We have tried to define the role, and the associated business, of Telcos in this framework. Telcos are in a good position to acquire, by using grid technology in their own processes, the necessary knowledge to be able to sell services to the users of the future Grid Network. Other companies are acquiring this knowledge: traditional hardware vendors and IT service providers are developing or implementing current grid technology solutions for their customers. Telcos should follow their example to gain a good position in the future *grid aggregator market.*

Finally, we speculated about a future in which all of us, connected from our homes or through our mobile devices, are users of grid technology. Some have sketched the ideas of playing games using grid platforms,[6] or selling the unused power of our devices to Grid Service providers, just as solar energy users sell the unused power back to the energy grid. This pervasive vision of the grid is still at least a decade away, but when it becomes a reality, telecommunications companies will have the opportunity to be some of the most relevant players in it.

---

[6] *http://www.butterfly.net/*

19 Grids in Other Industries

Ahmar Abbas

Grid Technology Partners

In This Chapter

- Grid Computing Adoption Track
- Examples of Grid Computing Application Areas

To: Grid Technology Partners
From: Vice President, Fixed Income Technology
    Big Wall Street Investment Bank
Subject: RE: Thoughts on grid computing

... trading desk managers used to looking at their previous day's closing positions now want to take advantage of the huge leaps in processor speeds to re-compute their desk's risk assessment at the drop of a hat. Additionally huge scandals with rogue traders being able to get away for days unnoticed, makes this a worthwhile exercise. To be able to pull up their entire desks exposures in a matter of minutes rather than tens of hours will become the norm rather than the exception. This appetite for risk computing power is hard to satiate, the demand leading the supply by leaps and bounds given the economy and the stakes at large. Companies are already turning heavily to idle computing power, leveraging desktop and other resources. The risk managers at the bank are relentless in their demands ...

19.1 INTRODUCTION

Grid Computing adoption and market development in enterprises will occur in three phases. We are currently in the very first phase of Grid Computing adoption. In this phase, the adoption is being lead by the product development and R&D groups within corporations. These groups are deploying Grid Computing to optimize product development and life cycle management. Many of the applications used in product development exhibit excellent parallel efficiency and are easily portable for use on grids. Grid deployments will range from high-performance grids to desktop grids. Utility grids to service this market are already being established. Each utility grid will be specialized in nature servicing specific applications.

The second phase of Grid Computing deployment will involve the *traditional IT* applications. These are business activity and process-related applications. The timeframe for this phase depends on the interests shown by the key software providers in this space. So far, the only commitment to grid-enable its products has come from IBM. BEA, Seibel, Peoplesoft, SAP, and others have been uncharacteristically quiet. Open Grid Service Architecture will drive the second phase of Grid Computing adoption.

The third phase of Grid Computing will involve grid-enabled applications that are available to consumers. Sony announced earlier this year that its next generation of its popular Playstation gaming system will be built using grid technology.

This chapter discusses some of the key sectors where Grid Computing is making a substantial impact.

19.2 GRIDS IN FINANCIAL SERVICES

Financial institutions are broadly defined to include commercial and merchant banks, investment and trading houses as well as insurance firms. Financial firms combined spend more per revenue dollar on their information technology infrastructure than any other industry.

Financial applications are challenged along two dimensions. First, they must process information, such as a portfolio risk model, before the information loses its value and; second, the modeling has to be performed with a degree of rigor that generates credible and actionable results.

Almost all financial modeling applications today are based on Monte Carlo simulations.[1] Monte Carlo applications are embarrassingly parallel, but the quality

---

[1] Monte Carlo simulations are not limited to financial applications, but are used in a wide spectrum of application areas such as advertising, marketing, forecasting, etc.

BOX 19.1 Grids requirements for Wall Street.

To: Grid Technology Partners
From: Vice President, Fixed Income Technology
Big Wall Street Investment Bank
Subject: RE: Thoughts on grid computing

... trading desk managers used to looking at their previous day's closing positions now want to take advantage of the huge leaps in processor speeds to re-compute their desk's risk assessment at the drop of a hat. Additionally huge scandals with rogue traders being able to get away for days unnoticed, makes this a worthwhile exercise. To be able to pull up their entire desks exposures in a matter of minutes rather than tens of hours will become the norm rather than the exception. This appetite for risk computing power is hard to satiate, the demand leading the supply by leaps and bounds given the economy and the stakes at large. Companies are already turning heavily to idle computing power, leveraging desktop and other resources. The risk managers at the bank are relentless in their demands ...

of the results is heavily predicated on the quality of random number streams employed by the application.[2] Scalable parallel random number generators increase the availability of statistically independent streams of random numbers and increase the confidence associated with the results. Therefore, Monte Carlo simulations are a good fit for Grid Computing deployment.

Unlike life sciences applications, many of which are in the public domain, most of the financial applications are primarily written in-house and are specific to each institution. These applications, however, can be easily ported to the grid environment with different parallelization tools, SDKs, and API.

Grid Systems of Mallorca, Spain has grid-enabled many proprietary Microsoft® Excel®-based applications for customers in the financial sector. In each of these cases, the client still used the Excel front end, except that each time they ran their portfolio volatility models, the processing was distributed to the laptops and personal computers at the office. The processing time, of course, was substantially reduced.

Additionally Grid Computing is being used at financial services institutions for portfolio modeling, stochastic valuation reporting, and asset liability management to speed analysis and improve critical decision making.

---

[2] Basney, J. and Livny, M., "High Throughput Monte Carlo," Computer Science Department, University of Wisconsin.

19.3 GEO SCIENCES

Overcoming geological uncertainty has been a constant challenge for companies in the oil and gas sector. Generally the oil and gas exploration business consists of four major phases. During the first phase, known as *basin analysis*, activities are carried out to decide whether to enter an exploration activity. In the second phase, known as *exploration*, activities are aimed at finding possible hydrocarbon accumulations and at evaluating the probability of being economically viable. During the *appraisal* phase, activities are carried to assess the potential and the characteristics of the hydrocarbon reservoir discovered in the exploration phase. In the *development* phase, the activities involving design and building of the facilities required to extract the hydrocarbons are concluded, after which the productions of oil or gas begins at the wells.

In each of the phases, a tremendous amount of data is collected and processed. In fact, there has been an exponential increase in the amount of information collected and processed from 300 MB per square km in the early 1990s to 25 GB per square km collected today.[3] On-shore exploration costs are close to US $ 20M per well and US $ 80M per well for off-shore drilling. Additionally the cost of seismic surveys, etc. can be up to US $35,000 per square kilometer.[4]

Grid Computing has played a major role in helping the oil and gas industry process the data efficiently and pinpoint suitable areas for drilling. In each of the phases, companies have had to compromise between the resolution of data collected and the time it took to process it. Grid Computing saves companies millions of dollars by allowing them to not only collect and analyze high-resolution data to pinpoint drilling sites, but also to do it with great speed. As indicated earlier, the cost of prospecting at the wrong location is quite substantial. Many of the oil and gas companies today are either replacing SMP/cluster-based solutions or complementing them with grid deployments.

Companies performing reservoir modeling, 2D & 3D seismic processing, and horizontal drilling benefit from Grid Computing's ability to increase scope and improve accuracy of data analysis.

19.4 MANUFACTURING

There are numerous applications in the manufacturing sector that are taking advantage of Grid Computing today. Large automotive manufactures are using Grid Computing for crash test simulation experiments, while aircraft manufactures are using it for simulating wind tunnels. *Computational fluid dynamics* (CFD)

---

[3] Luigi Salvador, *High Performance Computing for the Oil and Gas Industry*, 2003.

[4] ML Geovision, *www.alkorinternational.com/index.html*

is one area that is greatly benefiting from the power of Grid Computing. Computational fluid dynamics predicts fluid flow behavior, transfer of heat, mass, phase changes, chemical reactions, mechanical movement, and stress or deformation of solid structures. CFD analysis can predict various problems related to any of the above behaviors prior to product design. Hundreds of millions of dollars can be saved in prototype development costs and potential product delays by employing computational models to study system behavior under the various conditions listed above instead of discovering them later in the production stage. CFD is compressing product development time in numerous industries.

A family of CFD programs has been developed by various corporations and academia. Commercial CFD tools developed by Fluent Corporation today are extremely popular and widely deployed. These tools are currently deployed on cluster grids and possess the parallel efficiency that makes them suitable for desktop grids as well. The applications of computational fluid dynamics are seen in almost all major industries. Some specific applications are listed in the Table 19.1.

TABLE 19.1 CFD Application by Industry

Industry [1]	CFD Application [2]
Oil and Gas	Modeling drag forces for better oil rig platform design
Aerospace & Defense	Design of anti-icing systems
Automotive	Aerodynamics; fuel systems; air handling systems; lighting; engine cooling
Biomedical	Simulation of Blood Pump
Chemical Process	Pump design; mixing
Environmental	Aeration; containment dikes
Power Generation	Exhaust; duct flows
Steel	Combustion; oxygen steel making
Semiconductors	Crystal growth; ventilation

[1] Additional industries include Appliances, Electronics, Glass, HVAC&R, Marine and Off-shore, Polymer Processing, Turbomachinery

[2] There are probably hundreds if not thousands of applications of CFD in aerospace industries, the scope of which goes beyond the purpose of this report.

Source: Grid Technology Partners

19.5 ELECTRONIC DESIGN AUTOMATION

In EDA, the fundamental challenge stems from incremental progress into deeper sub-micron design technologies; this advance implies staggering challenges for design synthesis, verification, timing closure, and power consumption. Through direct association with Moore's Law, design synthesis has a gained profile. However, it is design verification that has an even greater potential to become the ultimate design bottleneck: As design complexity increases, verification requirements escalate rapidly. Section 6.3.1 contains a detailed discussion of a grid-based EDA application.

All the major EDA vendors—such as Cadence Design Systems, Synopsys, Inc., and Mentor Graphics—are active users of the technology. Many offer grid-enabled versions of their products as well.

19.6 ENTERTAINMENT AND MEDIA

Graphics processing, rendering, digital content compression, and encoding are some of the areas within the entertainment and media sector where Grid Computing is having an impact. Large as well as small studios are taking advantage of the technology to cope with the ever-increasing digital content of movies. An example of digital encoding is discussed in detail in Chapter 12.

Another focus is on digital content distribution. Broadcasters would like to share distributed media files and other distributed technical resources—building on technical solutions already developed for computing grids. For example, the British Broadcasting Corporation's (BBC) network faces many of the problems which computing grids are designed to solve. It is a distributed network with processing of broadcast material taking place at many nodes, carrying material (broadcast video) which has very high bandwidth requirements and mixes "live" and stored events. The BBC also has high demands for reliable and consistent performance from its systems and networks. Special purpose broadcast processing equipment (e.g., video editing suites or image rendering devices) are located at specific points in the network but, in general, are not available for use outside these locations. As an example, a producer from BBC Northern Ireland has to travel to London with the program material should he/she want to edit the program on an advanced editing suite not available at the BBC's Belfast site.

Until now, broadcast video has largely been stored as physical objects (tapes on shelves) and distributed using special-purpose machines via dedicated broadcast networks. However, the advance of information technology is now at a stage where video is being stored more and more as data files on servers and distributed using the IP networks familiar to the computing world.

The BBC has initiated a project that will investigate the application of technology developed for sharing distributed data files and computing resources in a computing environment to the broadcast world. If successful, it will allow the BBC to plan for a future broadcasting environment where broadcast material and dedicated broadcast resources can be shared much more easily in the BBC's distributed environment.

In this project, a baseline media grid will be developed that will manage the distribution of program content between a collection of distributed sites. The grid will enable program content distribution to implement a broadcast schedule, manage the security issues relating to distributed collections of stored program content, and demonstrate the integration of a selected broadcast technical resource in a grid environment.

19.7 CHEMICAL AND MATERIAL SCIENCES

Materials scientists, chemists and biochemists, physicists, and chemical engineers performing simulation and modeling can benefit from increased compute power for faster processing. Computational chemists, for example, are busy building property and structure databases for compounds and materials. Combinatorial statistics will be applied to these models to create and discover new compounds and materials that exhibit desired characteristics. Data grids are being used to address the volume of data (data, image, videos) while computational grids are being used for analysis and simulations.

19.8 GAMING

Massively multi-player games have become tremendously popular on the Internet over the past few years. NC Soft claims that over a million people pay to play its MMG Lineage™ over the Internet. Like digital rendering and animation, creating MMG games and then hosting them to be able to handle hundreds of thousands of users requires tremendous computational capacity. Grid Computing technology is being used today to create a scalable infrastructure to host MMG games. Electronic Arts has partnered with Oracle to build a grid for MMG games.

19.9 CONCLUSION

Grid Computing is making an impact across numerous industries. The list of applications that are taking advantage of Grid Computing is also expanding at a feverish pace. The availability of tools to grid-enable applications (Chapters 11 and 12) and easier integration with grid middleware (Chapter 13) has been extremely successful in bringing new applications to the grid.

20

Hive Computing for Transaction Processing Grids

Chris O'Leary

Tsunami Research

In This Chapter

- Introducing Hive Computing
- Building Transaction Grids with Hive Computing

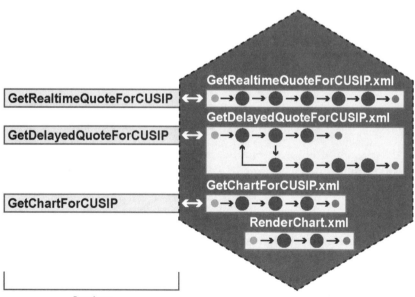

20.1 INTRODUCTION

The promise of Grid Computing—organizations being able to acquire all the power they need for only as long as is necessary—is incredibly compelling. Although grid computing has experienced significant success in bringing productivity gains and cost savings to engineering and scientific applications, it has not yet proven its mettle for highly transaction-oriented applications such as:

- Enterprise Resource Planning (ERP)
- Customer Relationship Management (CRM)
- Chain Management (SCM)
- E-commerce transaction and payment processing

This chapter makes the case that the current vision of Grid Computing is at least a bit fuzzy—and in some places largely incomplete—when it comes to questions of mission-critical computing in general and transaction processing in particular. This chapter discusses a new approach to the development, deployment, and management of mission-critical applications—called *Hive Computing*—that is designed to complement and extend the vision of Grid Computing.

Hive Computing enables businesses to build a transactional resource, called a Hive™, that can be plugged into a grid and host the transaction-oriented applications upon which businesses depend.

The goal of Hive Computing is to expand the range of problems that can be solved with a grid and bring the benefits of Grid Computing to the mainstream of business computing.

20.2 HIVE COMPUTING

Hive Computing defines a new type of resource called a *Transactional Resource* that can be integrated into an existing grid. The transactional resource handles all the transaction-oriented applications.

When it comes to transaction processing, the biggest problem businesses face today is the need to choose between reliability and affordability. Existing transaction processing solutions such as fault tolerant systems are reliable but depend on complex and expensive hardware. In contrast, commodity computers are affordable but are not reliable enough to do important work.

As a result, many businesses—and in particular mid-sized organizations—are forced to either spend more than they can afford or settle for lower levels of reliability than they would like.

Hive Computing's objective is to make system reliability affordable.

20.2.1 Assumptions

Before getting into the details of Hive Computing, it is important to first make the point that Hive Computing is based on a new set of assumptions. At first glance, these assumptions would seem to be of only peripheral interest. However, the history of innovation makes it clear that—because of the impact assumptions have on the design and implementation of a system—it is impossible to achieve order of magnitude advances by making incremental improvements to an existing system. Instead, as the designers of the steamship, the jet engine, and Transmission Control Protocol (TCP) found, the only way to achieve order of magnitude advances is to start off with a fresh set of assumptions. When it comes to transaction processing, three assumptions force businesses to choose between reliability and affordability:

- Everything counts
- Computers are reliable
- Computers are precious

As a result, when developing the idea of Hive Computing, Tsunami Research realized that new assumptions would be required if it was to make reliability affordable.

The key assumptions that lie behind Hive Computing are:

- The application is all that matters
- Failure happens
- Computers are disposable

20.2.1.1 The Application Is All that Matters

When it comes down to it, the only thing people care about is the application; the process that does the work and gets the job done. Contemporary solutions typically force developers to understand and worry about what is going on in the layers below the application, which increases the cost and complexity of developing and managing applications.

As a result, the first assumption that lies behind Hive Computing is that the application is all that matters. Developers should be allowed to focus on the task at hand—the application—and the environment should handle everything else.

20.2.1.2 Failure Happens

Various computer companies have spent the past 40 years trying to build a completely reliable computer. They have yet to succeed. Hive Computing assumes that failure is the rule, not the exception. Computers and other hardware are vulnerable to numerous technical and human-induced flaws and catastrophic failures. Hive Computing takes the lead from the designers of TCP in assuming that computers are unreliable, and designs systems that can deal with rather than fear system failure.

20.2.1.3 Computers Are Disposable

The third assumption that lies behind Hive Computing is that the skyrocketing price/performance ratio of commodity computers signals that they are becoming disposable. As a result, it no longer makes sense to treat computers as discrete entities. Instead, it makes far more sense, and is ultimately far less expensive, to treat pools of computers as a single resource that is communicated with and administered as a whole, not in parts.

20.2.2 Overview

Hive Computing is a comprehensive and integrated approach to the development, deployment, and management of transaction-oriented applications. The goal of Hive Computing is to enable businesses and other organizations to leverage the rising performance and falling prices of commodity computers to construct mission-critical computing solutions that are both reliable and affordable. In the world of Hive Computing, all work is performed by a system called a *Hive* (see Figure 20.1).

At the most basic level, a Hive does two things. First, a Hive receives and processes *requests* that are sent to it by a *client*. This client may be a standalone application, a point of sale terminal, or an application server. These requests ask the Hive to perform some *service* using the data contained in the request. These services could include processing a credit card or other transaction, analyzing some piece of biological data, or processing an order. A Hive also receives and processes requests that are sent to it by an *administrator*. This may be a request to deploy an updated version of the software that it is running in the Hive or to make a change to the configuration of the Hive.

Regardless of whether a request is generated by a client or by an administrator, all requests are broadcasted to the Hive as a whole, not to any of the individual *workers* that make up the Hive (see Figure 20.2). Upon receiving a request, the

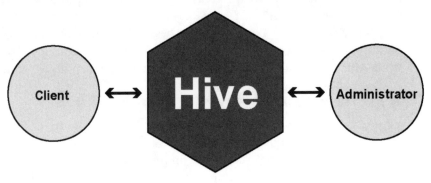

FIGURE 20.1 High level architecture.

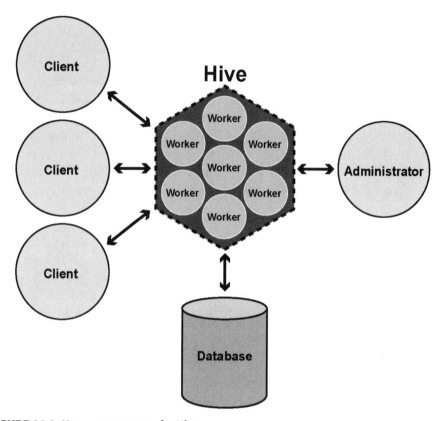

FIGURE 20.2 Key components of a Hive.

Hive as a whole decides which individual worker should handle the request and then hands the request to a worker.

A worker is an individual PC, a server blade, or some other computing resource, and has three characteristics. First, all workers must be located on the same logical network, although they do not need to be physically co-located. Second, regardless of their exact hardware configuration, the same software is deployed on all workers. Third, all workers are dedicated to the Hive. They cannot be used to perform some secondary function. Once an individual worker is assigned a request, it examines the request, determines which service it should carry out, and loads the code that is required to perform that service. Once the service has been performed, the worker returns the request to the client. At the same time that these actions are being performed by that worker, the Hive is continually monitoring its status to ensure that it performs the service.

20.2.3 **Capabilities**

This section outlines the capabilities of a Hive Computing-based infrastructure.

20.2.3.1 A Hive Is Self-organizing, Self-healing, and Self-managing

Hive Computing enables a Hive to automatically bring new workers online, work around the failure of individual workers, and ensure that all workers have the correct versions of the software that is running in the Hive. To deploy an application in a Hive, all the developer has to do is submit it to the Hive as a whole. From that point on, the Hive takes over the responsibility of making sure that all workers are updated and dealing with the failure of any individual worker(s).

20.2.3.2 A Hive Creates a Mission-critical Computing Environment

Hive Computing endows a Hive with a sense of time. When deploying services into a Hive, developers are able to give the Hive a sense of how long it should take to perform a service. This enables a Hive to monitor the processing of all requests and determine whether a worker is taking too long. If the Hive determines that a worker is taking too long, it can ask a different worker to take over the processing of that request.

20.2.3.3 A Hive Utilizes Large Numbers of Dedicated Commodity Computers

Hive Computing addresses this issue by enabling businesses and other organizations to build reliable computing resources using cheap, unreliable commodity computers.

Three things make this both feasible and economical. First, a Hive is made up of large numbers of workers; this may range from tens to hundreds or even thousands. As a result, the reliability—or lack thereof—of any individual worker is not an issue. Second, a Hive is built from dedicated resources. No worker can perform any other function. While this increases the acquisition cost of the system, it decreases the operational cost by enabling the Hive to maintain the configuration of each worker. This shift of costs from people to hardware makes sense given that people are getting increasingly expensive while hardware is getting cheaper every day.

Finally, although a Hive is made up of large numbers of dedicated resources, it is addressed and administered as a whole, not in parts. All communication with and administration of a Hive is done with the Hive as a whole.

20.2.3.4 A Hive Is Designed to Host Transaction-oriented Applications

A final point to make about Hive Computing is that, because it is designed to host transaction-oriented applications, it requires that applications be architected in a compatible manner. This means applications must be divided up into discrete components with clear, typically service-based interfaces. This is necessary to enable the Hive to be able to deliver reliability at a very granular level. This requirement has two implications.

First, it means applications have to be modified in order to run in a Hive. Second, it means that most packaged applications are not able to run as-is within a Hive. Although this will be an issue for companies with large amounts of legacy code, it will not be a problem for organizations whose applications are written in a modern, modular fashion.

20.2.4 Programming Model

Even though the operations of a Hive are driven by a very different set of assumptions, the process of defining services and deploying them in a Hive is designed to be both familiar and simpler. Any developer who has worked with J2EE, servlets, stateless session beans, Web Services, application servers, or transaction processing monitors will quickly become comfortable with the process of creating a service and deploying it in a Hive.

20.2.4.1 Services

At the core of the Hive Computing programming model is the concept of a *service*. A service is a discrete function that any worker and thus—as every worker is configured identically—the Hive as a whole can perform. Collectively, the services that a Hive can perform make up its *interface*. In the case of the Hive that is shown in Figure 20.3, it can perform the following services:

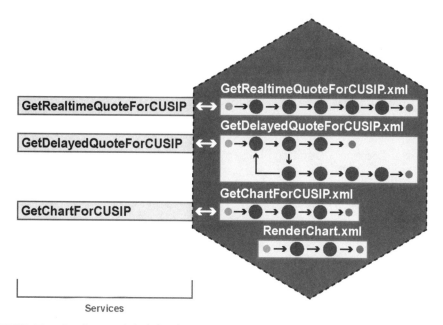

FIGURE 20.3 Services and their implementation.

- Get a real-time quote based on a CUSIP (stock identifier)
- Get a delayed quote based on a CUSIP
- Generate a 30-day or other price chart based on a CUSIP

In each case, the client submits a request to the Hive asking it to perform some service. This request is received by the Hive and is handed off to an individual worker for processing. Upon receiving a request, the worker finds the *process flow* that corresponds to that service. This process flow tells the worker what *tasks* to call and in what sequence, what to do depending on the results of executing each task, and what to do in the case of an error. Once the worker finishes executing all tasks, the worker then returns the request to the client.

20.2.4.2 Defining a Service

The process used to define a service is similar to that used in structured application development. In the case of structured application development, the developer first divides the code up into subroutines. Once these subroutines have been created, the developer then writes a master routine that calls these subroutines in order. This master routine controls the execution of the application based on the results that come out of these subroutines. When defining an individual service—or moving an existing application to a Hive—a developer would follow a similar process.

20.2.4.2.1 Tasks

The first step in the process is to define what are referred to as *tasks*. A task is the equivalent of a subroutine, function, or class and is smallest unit of work within a service. Similar to a stateless session bean, a task is a stateless, atomic component or other block of code—written in C, C++, or Java—that gets one or more pieces of data from a request that was sent in by a client, performs some function on or using that data, sets some value on the request, and hands the request back to the Hive. A task could perform a function such as:

- Looking up an item in inventory
- Purchasing an item
- Releasing the hold on an item

20.2.4.2.2 Process Flows

In a service, the equivalent of the main function or master routine is a process flow (see Figure 20.4).

A process flow is an XML document that allows the developer to declare to the Hive the sequence in which tasks are to be called, what to do based on different re-

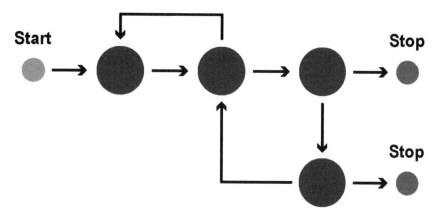

FIGURE 20.4 A simple process flow.

sults, and what to do if an error occurs. There are a number of advantages of using a process flow—and a declarative programming model in general—to perform this function. First, it allows developers to quickly and easily change definition of a service. As no coding or compiling is required, this makes a Hive far more adaptable. Second, it encourages the reuse of components. Because any task can be called from any process flow, the need to create duplicate code is reduced. This has the added benefit of allowing novice developers to utilize tasks that have been written by more experienced members of the organization. Finally, it allows developers to ask the Hive to perform functions like checkpoints, again without having to write any code. Developers can set attributes within a process flow and instructs the Hive to backup the state of a request. This allows a Hive to recover from the loss of an individual worker without having to start over from the beginning of a request. As Figure 20.5 demonstrates, whenever a request comes into a Hive, it is received by some worker (in this case, Worker A). This worker then selects a second worker (in this case, Worker B) to process that request. If at some point Worker B stops functioning, the Hive will notice this and will instruct Worker C to take over the processing of the request. Worker C is able to pick up exactly where Worker B left off because the modifications that the first worker made to the request are always backed up on another worker.

20.2.4.3 Deploying a Service

Once all of the required tasks and process flows that underlie a service have been created, they are then put into a *package*. That package is then deployed to the Hive as a whole, not to any individual worker. Upon receiving this package, the Hive assumes the responsibility for distributing these new tasks and process flows to each

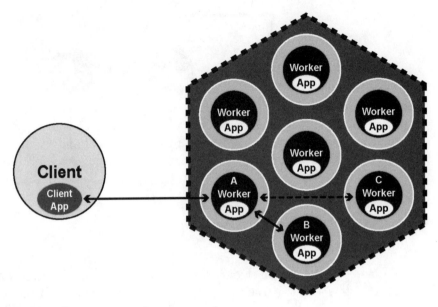

FIGURE 20.5 The request-service-response loop.

worker. Importantly, at no point does a Hive have to be stopped or bounced in order to deploy a new service. Instead, a new service can be deployed into—and an existing service can be updated in—a running Hive.

20.2.5 Benefits

These capabilities enable Hive Computing to deliver both reliability and affordability and give it a number of advantages over existing mission-critical computing solutions.

20.2.5.1 Reliability

When it comes to Hive Computing, reliability is broadly defined as the ability to handle any load and survive—and then work around—any failure. In particular, that means a reliable system is one that delivers:

- Scalability
- Availability
- Predictability

20.2.5.1.1 Scalability

Systems such as mainframes, fault tolerant computers, application servers, and operating system-based clusters, which are often used for transaction processing, are nominally scalable. However, in each case that scalability is significantly limited.

In the case of mainframes and fault tolerant computers, increases in capacity are generally made in large, expensive, and infrequent leaps due to the time and cost of acquiring and configuring new systems. This pattern also tends to hold for application servers because, while they can theoretically be deployed on large numbers of less expensive boxes, the configurations of all of those boxes would have to be maintained by hand. As a result, most application servers tend to be deployed on small numbers of powerful, and expensive, servers. Operating system-based clusters share the same problem as application servers. Although they can be deployed on commodity computers, all of those computers must be maintained by hand. As a result, most are either quite small or are maintained by teams of system administrators. In order to keep businesses from having to deal with these kinds of problems, a Hive is designed to be able to grow as needed but in a manner that is both more incremental and more affordable. Workers can be added in small, affordable increments and over an extended period of time. More importantly, this can be done without imposing significant maintenance costs on the system. All an administrator has to do to add a worker to a Hive is to plug it into the same network. From that point on, the Hive will Assimilate™ that worker and install all of the software it needs to start processing requests.

20.2.5.1.2 Availability

Traditionally, availability is achieved through the use of proprietary hardware. In this case, systems were built with dually redundant CPUs, memory, disk drives, power supplies, cooling fans, and other critical components. While this approach generally worked, it was both expensive and still vulnerable to catastrophic failure. In order to drive down the cost of availability, a Hive depends on software, not hardware. All of the important work in a Hive is performed by the Hive as a whole, not by any individual machine. This absence of a single point of failure means that a Hive can be built using cheap, commodity computers but still deliver the availability of a conventional fault tolerant system.

20.2.5.1.3 Predictability

Because the Grid Computing model relies on borrowed resources, it is hard for a grid-based application to make very many demands on the Compute Resource on which it is running. Instead, it must be prepared to get out of the way if that Compute Resource has to perform some other, higher priority task. Although this is often an acceptable tradeoff, it will not work when it comes to transaction processing. As a result, a Hive possesses a sense of time and is built from dedicated resources. Thus, a Hive is able to guarantee that a service will be performed within a given time period.

20.2.5.2 Affordability

When it comes to Hive Computing, affordability is broadly defined as a low Total Cost of Ownership (TCO). In other words that means low acquisition, development, operational, and maintenance costs. In particular, that means an affordable system is one that delivers

- Usability
- Adaptability
- Maintainability
- Commodity Components

20.3.5.2.1 Usability

Conventional approaches to the development of transaction processing applications assume that the developer understands—and is responsible for—all of the layers of the system. That means the hardware, the operating system, the storage mechanisms, the security infrastructure, and the application that sits on top of those layers. Although this assumption may have been valid in the past, it is becoming more and more problematic because people who understand all of these layers—and can deal with the resulting complexity—are both scarce and expensive. In order to solve this problem, a Hive is designed to be interacted with as a single, logical entity. At no point does the developer or administrator have to worry about individual workers. The Hive takes care of that. This yields two main benefits:

First, it means that developers are able to focus on the application, not the infrastructure, and are able to quickly and easily create reliable applications. Second, it makes plugging a Hive into a grid a straightforward process.

When it comes to writing an application, developers do not have to worry about handling issues such as load balancing and failover. The Hive will handle that. Similarly, deploying that application is a simple process. All the developer or administrator has to do is submit the application to the Hive. From that point on, the Hive will take over the responsibility of ensuring that all workers are updated.

20.3.5.2.2 Adaptability

One of the problems that businesses run into when building transaction processing systems is that they tend to be quite brittle; small changes can cause them to break. As a result, once such systems have been implemented, they become relatively rigid because people are afraid of changing anything and thus breaking something. In contrast, a Hive is quite adaptable. If new workers need to be added to a Hive, they must simply be plugged into the network. From that point on the Hive will Assimilate the machine and enable it to join the Hive as a new worker. In the same way, if a worker needs to be shifted from one Hive to another, the administrator only has to issue a single command to the Hive as a whole. A given number of workers will then drop out of that Hive and be Assimilated into the second Hive. Because of this ability to easily move workers from one Hive to another, organizations will be able to move capacity around as it is needed. In some cases, organizations may choose to make those shifts on a daily or even hourly basis.

20.3.5.2.3 Maintainability

One thing that keeps Grid Computing—which was designed largely from the point of view of academia—from being able to deliver the same value in the business world is that the economics of the two worlds are very different. In the world of academia, hardware is expensive and people are less so. Most universities and research institutions have access to huge pools of inexpensive, curious, and talented people, e.g., graduate students, post-docs, and interns. On the other hand, contemporary hardware can be hard to come by quickly (if at all). As a result, systems that come out the world of academia tend to be labor intensive and designed to be used on a shared resource. In contrast, in the world of business, people are expensive and hardware is inexpensive (or at least available). Most businesses spend tremendous amounts of time and money trying to find qualified people. In contrast, if they need to buy a piece of hardware, they can generally get it if they can make a good business case. As a result, businesses tend to be most interested in systems that allow them to do more with fewer people, not fewer resources. Businesses tend to find that it is better to waste the time of machines, not people. As a result, a Hive is self-managing and is able to perform functions such as Assimilation. This makes a Hive extremely easy to maintain and drives down its TCO.

20.3.5.2.4 Commodity Components

Although we live in an age where extremely powerful computers are remarkably inexpensive and plentiful—and in many cases virtually disposable—much of our thinking still reflects the thinking of the 1960s, 1970s, and 1980s when computers were expensive and scarce. For example, companies still talk about enabling organizations to get more from their "underutilized resources." The problem is that this attitude, which may still make sense in the world of academia, increasingly makes less and less sense in the world of business. Spend any amount of time studying the history of the computer industry, and one pattern you will notice is that many of the most significant advances were preceded by the decision to stop regarding a particular resource as precious and instead regard it as disposable. This happened in the 1980s with transistors and in the 1990s with microprocessors. Growing evidence—such as the rise of blade servers—indicates that this is on the verge of happening with PCs. Their prices have fallen to such a degree that in many cases it makes the most sense to regard them as disposable.

As a result, Hive Computing is designed to make it economical for businesses to take advantage of the rising performance and falling prices of commodities.

20.4 CONCLUSION

The ease with which highly reliable applications can be developed for, and deployed into, a Hive enables the value of Grid Computing to be extended in a number of ways. Just as Grid Computing enables academic and research organizations to create computational grids, Hive Computing enables businesses to create transactional grids (see Figure 20.6). By plugging one or more Hives into a grid, businesses—and research institutions—will be able to gain access to a resource that provides them the transactional capabilities they require to perform their work in the most reliable and efficient manner. More important, the self-organizing, self-healing, and self-managing capabilities of a grid mean that these transactional capabilities can be acquired at a low cost. All a user has to do is create a package and give it to the Hive. From that point, the Hive will take over the responsibility for ensuring that all

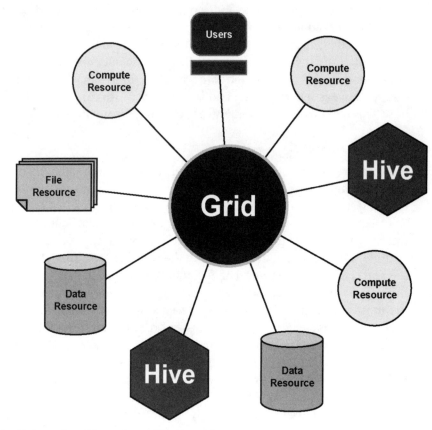

FIGURE 20.6 Two Hives plugged into a grid.

workers are updated. These capabilities also make it easy for a Hive to switch from providing one set of services to providing a different set of services. As a result, this will create opportunities for organizations to create both intra-company and inter-company resource pools.

These resource pools will be collections of workers that can be quickly and easily moved from Hive to Hive as the needs of their users change. For example, businesses may use these resource pools to expand the size of a Hive in order to handle a seasonal or other periodic increase in demand.

Regardless of precisely how businesses choose to leverage its power, there is no doubt that—by building on and extending the idea of Grid Computing—Hive Computing will provide businesses and other organizations with tremendous levels of flexibility and drive down the cost of reliability.

A

About the CD-ROM

This CD-ROM contains numerous tools you can use to get started with Grid Computing. There are three primary folders: Engineered Intelligence CxC, Grid-Iron Software XLR8 v1.2 and United Devices Grid Calculator. The contents of these folders are listed below.

To use the items on the CD-ROM, you should select the folder you wish to use and then select the appropriate files. Again, be sure that you have all the necessary hardware and software to run these files.

■ Engineered Intelligence CxC: Engineered Intelligence (*www.engineeredintelligence.com*) has provided their CxC parallel computing development system. Detailed documentation on the system and a programming guide is also included.

■ GridIron Software XLR8 v1.2: For software developers with computationally intensive applications, GridIron XLR8™ is an application development tool and runtime infrastructure that makes it Profoundly Simple™ to develop, use and manage applications with the added speed of distributed computing. XLR8 enables your computationally intensive software application to run faster on multiple computers. This CD includes copies of GridIron XLR8™ version 1.2 for the Windows® and Mac® OS X operating systems. This product is fully functional subject to the terms of the GRIDIRON DEVELOPER SOFTWARE LIMITED LICENSE AGREEMENT.

■ United Devices Grid Calculator: United Devices (*www.ud.com*) has provided a calculator that allows users to determine the aggregate computational capacity of their enterprise IT infrastructure. The Grid Calculator is supported on Windows 98 systems.

SYSTEM REQUIREMENTS:

The CD-ROM content can be accessed by any Microsoft Windows based system. However please refer to documentation for each individual software contribution for systems requirements for installation and execution.

Index

A

X

Y